Reihe der Villa Vigoni
Band 1

Deutsch-Italienisches Zentrum »Villa Vigoni«

Robert Zwilling Cesare Balduini (Eds.)

Biology of Aging

With 67 Figures
and 21 Tables

Springer-Verlag

Berlin Heidelberg New York
London Paris Tokyo
Hong Kong Barcelona
Budapest

»Villa Vigoni«
Via Giulio Vigoni 1, Loveno di Menaggio
I-22017 Menaggio (Como), Italy

Professor Dr. Robert Zwilling
Institute of Zoology, University of Heidelberg
Department of Physiology
Im Neuenheimer Feld 230, D-6900 Heidelberg 1 (W)
Federal Republic of Germany

Professor Dr. Cesare Balduini
Department of Biochemistry, University of Pavia
Via Bassi, 21, I-27100 Pavia
Italy

ISBN 3-540-54488-7 Springer-Verlag Berlin Heidelberg New York
ISBN 0-387-54488-7 Springer-Verlag New York Berlin Heidelberg

Library of Congress Cataloging-in-Publication Data
Biology of aging / Robert Zwilling, Cesare Balduini (eds.) p. cm. – (Reihe der Vigoni: Bd. 1)
Contains papers presented at a symposium held in Oct. 1990 at the Villa Vigoni on Lake Como, Italy.
Includes bibliographical references and index.
ISBN 3-540-54488-7 (alk. paper). – ISBN 0-387-5488-7 (alk. paper)
1. Aging – Physiological aspects – Congresses. I. Zwilling, Robert. II. Balduini, C. (Cesare) III.
Series: Reihe der Villa (Berlin, Germany): Bd. 1.
[DNLM: 1. Aging-physiology-congresses. WT 104 B6147 1990] QP86, B518 1992 612,6'7 – dc20
DNLM/DLC for Library of Congress 92-2165 CIP

Printed in Germany

Production Editor: Martin Langner
Typesetting: Mitterweger Werksatz, Plankstadt
25/3145 5 4 3 2 1 0 Printed on acid-free paper

Preface

This volume contains the papers presented at the symposium "Biology of Aging", held at the Villa Vigoni on Lake Como in October 1990. More than 40 researchers in the field attended the meeting, which was the result of cooperation between the Universities of Heidelberg and Pavia. It provided scientists and PhD students an opportunity to meet for 4 days to discuss a variety of aspects of one of the most exciting topics in modern biology. The symposium did not focus on one specific biological system or phenomena, but attempted to cover as many different aspects of the aging process as possible, especially those most prominently featuring in the ongoing research on aging. We hoped that this would further stimulate experimental research.

The 16 lectures presented at the meeting (eight each from participants from Heidelberg and from Pavia) dealt with age-related processes in different biological systems and at different levels of organization. Much attention was paid to aging of the nerve system, which was considered both with respect to its biochemical and physiological implications, as well as to the age-related pathological alterations. The reports on Alzheimer disease instructively combined these different aspects of nerve cell aging. While many papers focused on the age-dependent changes in DNA, protein, membranes, and cell organelles, the aging of connective tissue was also discussed, with attention being particularly paid to atherosclerotic degeneration.

The meeting was opened by an overview of the biology of aging, and closed by a consideration of the fundamental role of genetic control in the aging process.

We hope that this book may not only provide an up-to-date collection of important findings in the field of aging research but that it might at the same time encourage young scientists to become engaged in this exciting and promising scientific endeavor.

We gratefully acknowledge the most effective support by the Villa Vigoni Foundation, which made this symposium possible.

June 1991, Pavia and Heidelberg

Cesare Balduini
Robert Zwilling

Contents

Contributors

Balduini, C.
Dipartimento di Biochimica
Via Bassi, 21
I-27100 Pavia

Bellomo, G.
Dipartimento di Medicina Interna
e Terapia Medica
Clinica Medica I
Università di Pavia
I-27100 Pavia

Benedetti, A.
Istituto di Patologia Generale
Università di Siena
I-53100 Siena

Benevento, A.
Dipartimento Di Chirurgia
Università di Pavia
Ospedale di Circolo
Viale Borri
I-21100 Varese

Benzi, G.
Istituto di Farmacologia
Università di Pavia
Piazza Botta 11
I-27100 Pavia

Beyreuther, K.
Zentrum für Molekularbiologie
Universität Heidelberg
Im Neuenheimer Feld 282
D-6900 Heidelberg (W)

Biamonti, G.
Istituto di Genetica Biochimica
ed Evoluzionistica, CNR
Via Abbiategrasso 207
I-27100 Pavia

Brovelli, A.
Dipartimento di Biochimica
Via Bassi, 21
I-27100 Pavia

Bürkle, A.
Institut für Virusforschung/ATV
Deutsches Krebsforschungszentrum
Im Neuenheimer Feld 506
D-6900 Heidelberg (W)

Castellana, M. A.
Dipartimento di Biochimica
Via Bassi, 21
I-27100 Pavia

Costa, A.
Dipartimento di Neurologia III
Fondazione C. Mondino
Università di Pavia
I-27100 Pavia

Curti, D.
Istituto di Farmacologia
Università di Pavia
Piazza Botta 11
I-27100 Pavia

Cohen, S. A.
Zentralinstitut
für Seelische Gesundheit
Abteilung für Psychopharmakologie
D-6800 Mannheim (W)

DeLuca, G.
Dipartimento di Biochimica
Via Bassi, 21
I-27100 Pavia

DeRenzis, M. R.
Dipartimento di Biochimica
Via Bassi, 21
I-27100 Pavia

Dionigi, R.
Dipartimento di Chirurgia
Università di Pavia
Ospedale di Circolo
Viale Borri
I-21100 Varese

Focher, F.
Istituto di Genetica Biochimica
ed Evoluzionistica, CNR
Via Abbiategrasso 207
I-27100 Pavia

Fulceri, R.
Istituto di Patologia Generale
Università di Siena
I-53100 Siena

Genazzani, R.
Dipartimento di Ostetrica
e Ginecologia
Università di Modena
I-41100 Modena

Grube, K.
Institut für Virusforschung/ATV
Deutsches Krebsforschungszentrum
Im Neuenheimer Feld 506
D-6900 Heidelberg (W)

Gelbmann, C. M.
Zentralinstitut
für Seelische Gesundheit
Abteilung für Psychopharmakologie
D-6800 Mannheim (W)

Hoyer, S.
Abteilung für Pathochemie
und Allgemeine Neurochemie
Universität Heidelberg
Im Neuenheimer Feld 220/221
D-6900 Heidelberg (W)

Hartmann, H.
Zentralinstitut
für Seelische Gesundheit
Abteilung für Psychopharmakologie
D-6800 Mannheim (W)

König, G.
Zentrum für Molekularbiologie
(ZMBH)
Universität Heidelberg
Im Neuenheimer Feld 282
D-6900 Heidelberg (W)

Küpper, J.-H.
Institut für Virusforschung/ATB
Deutsches Krebsforschungszentrum
Im Neuenheimer Feld 506
D-6900 Heidelberg (W)

Masters, C. L.
Zentrum für Molekularbiologie
Universität Heidelberg
Im Neuenheimer Feld 282
D-6900 Heidelberg (W)

Martignoni, E.
Dipartimento di Neurologia III
Fondazione C. Mondino
Università di Pavia
I-27100 Pavia

Mazzarello, P.
Istituto di Genetica Biochimica
ed Evoluzionistica, CNR
Via Abbiategrasso 207
I-27100 Pavia

Metz, J.
Institut für Anatomie
und Zellbiologie
Universität Heidelberg
Im Neuenheimer Feld 307
D-6900 Heidelberg (W)

Minetti, G.
Dipartimento di Biochimica
Via Bassi, 21
I-27100 Pavia

Mirabelli, F.
Dipartimento di Medicina Interna
e Terapia Medica
Clinica Medica I
Università di Pavia
I-27100 Pavia

Mönning, U.
Zentrum für Molekularbiologie
Universität Heidelberg
Im Neuenheimer Feld 282
D-6900 Heidelberg (W)

Müller, W. E.
Zentralinstitut
für Seelische Gesundheit
Abteilung für Psychopharmakologie
D-6800 Mannheim (W)

Nappi, G.
Dipartimento di Neurologia III
Fondazione C. Mondino
Università di Pavia
I-27100 Pavia

Osiewacz, H. D.
Deutsches Krebsforschungszentrum
Institut für Virusforschung
Angewandte Tumorvirologie
Im Neuenheimer Feld 506
D-6900 Heidelberg (W)

Passi, A.
Dipartimento di Biochimica
Via Bassi, 21
I-27100 Pavia

Petraglia, F.
Dipartimento di Ostetrica
e Ginecologia
Università di Modena
I-41100 Modena

Polak, J. M.
Dipartimento Patologia Umana
Via Forlanini, 26
I-27100 Pavia

Piccinini, G.
Dipartimento di Biochimica
Via Bassi, 21
I-27100 Pavia

Prior, R.
Zentrum für Molekularbiologie
Universität Heidelberg
Im Neuenheimer Feld 282
D-6900 Heidelberg (W)

Rindi, G.
Dipartimento Patologia Umana
Via Forlanini, 26
I-27100 Pavia

Schubert, T.
Zentralinstitut für
Seelische Gesundheit
Abteilung für Psychopharmakologie
D-6800 Mannheim (W)

Seppi, C.
Dipartimento di Biochimica
Via Bassi, 21
I-27100 Pavia

Solcia, E.
Dipartimento Patologia Umana
Via Forlanini, 26
I-27100 Pavia

Spadari, S.
Istituto di Genetic Biochimica
ed Evolizionistica, CNR
Via Abbiategrasso 207
I-27100 Pavia

Stoll, L.
Zentralinstitut für Seelische
Gesundheit
Abteilung für Psychopharmakologie
D-6800 Mannheim (W)

Verri, A.
Istituto di Genetica Biochimica
ed Evoluzionistica, CNR
Via Abbiategrasso 207
I-27100 Pavia

Zwilling, R.
Zoologisches Institut
Universität Heidelberg
Im Neuenheimer Feld 230
D-6900 Heidelberg (W)

Aging – Still a Mystery

R. Zwilling

Aging is one of the most intriguing and least understood phenomenon in modern biology. Nonetheless, it is present everywhere around us – and it is in us. Man has always followed with amazed curiosity and emotional interest the changes accompanying birth, development, aging, and death. And human beings date back to the very day when they became aware of aging and death and started to bury the dead. But superstition and ignorance also have always obscured the occurrence of aging and Christoph Wilhelm Hufeland, who wrote the medical bestseller of the 18th century ("On the Art of Prolonging Human Live", 1796 [1], rightly stated:

> "This problem has always been a favorite of the keenest minds, a playground for daydreamers, and the main enticement for charlatans and swindlers."

In the meantime, many different and contradicting theories have been forwarded to explain aging, and the large number of these theories alone makes us realize that we still lack satisfactory insight into the central processes of the aging phenomenon.

It was August Weismann (born 1834 in Frankfurt/M., died 1914 in Freiburg i. Br.), Professor of Zoology, who in 1881 laid the foundations of our modern understanding of aging [2]. His central thought is that aging is not inherent in life itself but that it is a later addition due to evolution. He contends that, firstly, such a connection is hardly visible and, secondly, that death as a consequence of aging does not occur in all organisms.

Death through aging is not a necessary attribute of life. Weismann points to the example of a dividing amoeba and asks: does this division end an individual life and do two new individuals emerge from that?

Opposing objections for example, that after thousands of cell divisions only a correspondingly small amount of the original substance would be present and obviously the original cell would have perished, he asks: Where is the corpse? What is dying? Also, the individual amoeba does not disappear any more than the man who is no longer the individual he was as a boy 20 years ago.

From here Weismann concluded that death from inner causes could not be established in unicellular organisms, since both daughter cells have the same quality and hence both are the bearer of unbounded life. Unicellular individuals therefore die by accident (being devoured, by starvation, from en-

Reihe der Villa Vigoni, Band 1
Biology of Aging
Zwilling and Balduini (Eds.)

vironmental influence), not from age. Since all multicellular organisms are derived from monocellular ones, consequently for them there will have been no aging, in the beginning either.

Now, when and why did aging come into life? Pluricellular organisms arise when after cell division single cells stick together. But this cannot stay as it is: a clump of cells not only has no advantage over a single cell but rather great disadvantages, for example, severe problems in nutrition and excretion. The way out is differentiation and specialization.

Some cells begin to form cilia for locomotion, and others specialize for functions like digestion, excretion or signal transduction. But what about the primary function of reproduction? It is now confined to a cell type specialized for propagation, the germ cells. They necessarily have retained the old property of being immortal and form a cell line that guarantees the continuity of life.

Not so the remaining specialized somatic cells: whatever their specific function might be, it does not absolutely encompass permanence and they therefore can become mortal.

Weismann concludes: "In unicellular animals it was not possible to install normal death, because individual and reproductive cell are still one and the same. In multicellular organisms somatic and generative cells have separated, death by aging became possible and we see that it has been established indeed." Therefore only specialized somatic cells undergo a time-dependent, irreversible decline, while germ cells fuse and give rise to a new individual which possess germ cells of his own.

On these grounds a species can be defined as a continuous germ line which is propagated by an endless row of mortal individuals. In the course of evolution selection will prevent individuals, who have not adapted from procreating, and by this a gradual alteration of the germ line will occur, the whole species adapting to the changing internal and external conditions.

In this view the limited life span and the transitoriness of the individual are nothing less than a prerequisite for the permanence and the further development of the species.

Johann Wolfgang von Goethe says in his "*Schriften über die Natur*" [3]:

> "She [i.e. nature] has invested everything in individuality – and cares little about individuals."

We understand that aging has been the price for our own existence.

But, do all multicellular organisms age in the same way? A 3 000-years-old sequoia raises an interesting issue: it contains no living cell older than 30 years. And if we do not use dead wood to define age – which evidently makes no sense – how old is such a tree? It has been recognized that these huge trees consist of indefinitely dividing cell lines which, in principle, continue growing and do not age in the sense it has been defined before.

Their end is brought about by windfall or fire and a further increase in height is limited by the transport capacity to span the distance between roots and leaves.

Surprisingly, these largest and oldest living objects on earth do not age very differently from some of the tiniest and shortest-living organisms, the amoeba. Such observations are not limited to plants, but can be made in animals as well. In the fresh water polyp *Hydra* it is not the whole organism that ages but the individual cell lines of which it consists. *Hydra* dies continuously at its basal and apical ends and is steadily re-born by the immigration of new cells from the generative zone. This multicellular organism reflects a level in the evolution of the aging process where it is still possible to permanently regenerate the whole body.

If one experimentally prevents the cells in the generative zone from dividing, *Hydra* will also age gradually within a few weeks, and it will do so in the same measure as the remaining cell populations decline.

Annual plants exhibit yet a different aging pattern: in only one vegetation period they undergo a defined series of events from germination to death which is precisely programmed. There have been reported mutations which alter the time course of the development of leaves, the blooming, the formation of seeds and finally the withering of the whole plant. In clover, by mutation of a single gene out of an annual form a biennial variety can result, thus doubling the life span. Here for the first time we encounter the phenomenon that life span can be genetically programmed.

An interesting model for programmed aging is represented by *Caenorhabditis elegans*, a nematode which lives in the soil and is 0.2 mm long [4]. The adult organism consists of exactly 959 cells ("cell constancy"). The nerve system is correspondingly simple and is composed of 302 cells (while in humans there are 10^{12} and in *Drosophila* there are still 10^5 nerve cells present). In development six stem cells establish independent cell lines from which the structures of the body are formed. Each cell line has its own inner schedule, a specific endogenic rhythm of cell divisions that cannot be disturbed by exogenic influences. Now, *Caenorhabditis elegans* dies after only 3 weeks of old age. The question arises: how can 959 cells age in 21 days like clock work?

At least two findings point to the explanation that this is achieved by an exact genetic program. Already during embryogenesis cell death may be programmed: in a short time span of only 7 h precisely 113 young cells will die from inner causes during development.

This phenomenon is also known in vertebrate development, for example, when the pentadactyl extremities of birds and mammalians are shaped from an undifferentiated bladelike limb by extensive programmed cell death.

In development it seems to be less expensive to use primarily a common pattern and to introduce secondarily necessary specific variations by programmed cell death rather than to install a program for each such deviation.

The occurrence of such a phenomenon like programmed cell death demonstrates, however, that the organism is in a position to switch off a given cell at any time when so required by the program.

During aging, analogous processes might play a decisive role. In addition to that it was demonstrated that the fixed life span of *C. elegans* is accessible to experimental manipulation. By selective breeding a long-living mutant was created which lived for 31 days, corresponding to a 50 % increase in life span. This seems to confirm the theory that life span is under genetic control.

A biological clock also seems to determine human aging. We often gain surprising insight into the normal course of events when in those rare cases nature deviates from its original plan. Progeria, which causes death from premature senescence in 15-year old youths is such an incident [5]. The Werner syndrome (Progeria adultorum) was described in 1904 for the first time and since then about 200 cases have become known worldwide. A frequency of one in a million births has been calculated.

The first symptoms do not appear until the 2nd or 3rd decade of life when loss of hair, senile changes of the skin, cataracts, osteoporosis, arteriosclerosis and, in 50 % of the cases, an insulin-resistant diabetes mellitus can be observed. The syndrome is diagnosed at the age of 39 years on average and death occurs at an age of 47 years on the average. The mode of inheritance of the syndrome is autosomal-recessive and appears in some families with high frequency.

For the Hutchinson-Gilford syndrome (Progeria infantum) an autosomal-recessive hereditary disposition is presumed, too, even if some researchers believe that at least some cases must be explained by a dominant spontaneous mutation, also because the accumulation in certain families is less obvious. This syndrome was detected in 1896 and since then some 100 cases have been reported. The incidence is presumed to be 1 in 8 million births.

While at birth the children are apparently still normal, in the 3rd year of life they develop alterations of hair and skin.

They die at 12 – 18 years, often of a myocardial infarction, after having aquired many attributes of premature senescence.

Interestingly, fibroblast cells derived from progeria patients also in isolated cell culture do show premature aging.

The existence of the progeria syndrome can be best interpreted by the assumption that in humans, too, a genetically determined biological clock is precisely limiting the span of life and that in very rare cases this clock does not run correctly. Thus, the progeria syndrome is a derailment of the normal process of aging.

The simple fact alone that there is something like a discrete life span assigned to each mammalian species, but widely differing from each other, should tell us a lot about aging. No mouse has ever lived for 5 years but other mammals are still in their childhood at that age. Between the closely related mammals the span of life can vary by a factor of 50.

The time elapsed since the existence of the common ancestor of humans and chimpanzee was not sufficient to introduce a single amino acid substitution in the α- and β-chains of hemoglobin and in cytochrome c: these proteins, comprising together about 400 amino acid building blocks, are simply identical in man and in chimpanzee [6].

However, the life span of these two species varys by at least 100 %. This seems to indicate that both intelligence and longevity can easily be aquired during relatively short periods of evolution, while it is obviously much more troublesome to change ones hemoglobin.

Why then do some mammals live 50 times longer than others? Recently by mathematical analysis it was shown that the four factors brain weight, body weight, specific metabolic turnover (calories per gram) and body temperature account already for 85 % of the differences of life span observed. This does not, however, answer the question whether some species profit from a short and others from a long span of life, possibly dependent on their diverse conditions of life, or whether the life span of a species is relatively invariant to selective pressure.

Some believe that humans are growing older and older. In what sense is this true? For clarification we have to distinguish between life expectancy and life span.

In ancient Greece life expectancy was 22 years, but Sophokles lived for 90 years. Since then in the industrialized world the life expectancy has at least tripled, but the life span of man has probably not changed at all over the last million years.

While the example of Sophokles tells us that it has always been possible for some individuals to come close to the natural life span, we are approaching the state when this is within reach for virtually all of us.

While life expectancy has rapidly increased, life span has not. This poses an insurmountable barrier to modern medicine. It has been argued that the complete elimination of the three major causes of death – heart disease, arteriosclerosis and cancer – would only add a few years to our life expectancy because in the developed countries we are already in the range of the invincible biological span of life.

The gerontologist Alex Comfort contends:

> "Today, nearly all our intractable problems that are not psychosomatic are geriatric."

According to him the aging process is characterized by a steady loss of homoeostasis, i. e., a constant decline of vital physiological functions and capacities. This is especially evident in decreased immune response or reduced heart output. Most impressive are also the 50 % survival rates after severe skin burning. While a 5-year-old child will tollerate a 51 % loss of its skin, the area drops at 55 years to 23 % and at 70 years to only 9 %.

Most physicians agree that actually nobody dies of old age alone, but rather the gradual failure of homoeostasis, which results in the probability of dying from an ordinary disease doubling every 8 years. Therefore comfort has called aging "a self-aggravating process." It proceeds exponentially resulting in the sharply marked life spans we observe.

What causes the decline of homoeostasis? Many theories have been formulated to explain this. They all can be attached to two alternative, mutually exclusive lines of thinking.

The error catastrophe theory contends that during a lifetime there must be a steady loss of information and an accumulation of missense in all somatic cells due to mutations and the inevitable errors occurring during the copying of information [7]. There is no mechanism visible which counteracts this deterioration, as for example, for the germ line cells which are under the control of selection.

As a consequence this must lead to an increased synthesis of aberrant or defective proteins. Especially deleterious in this case would be the effect on enzymes which themselves are involved in information transfer, and this is true for enzymes which play an active role in transcription and in translation.

Such errors would then lead to a self-accelerating process of dysfunction that in many respects resembles the manifestations of aging that have been described above. This would end in an "error catastrophe" (Leslie Orgel [7]) for the organism, which we experience as aging and death.

It has indeed been observed that during aging the number of chromosomal aberrations significantly increases. According to Curtis in mammals in the second half of their life up to 70 % of all liver cells show abnormal chromosomes [8].

But, what bearing this might have on the aging process is by far not clear. Furthermore, the earlier reports on aberrant proteins and the effect they might have on the aging process have almost completely vanished in recent years.

Much of what has been said so far points, however, to the direction of the program theory. This theory is based on the assumption that the aging process is determined by a rigid genetic program and not by a statistical accumulation of errors. Every higher organism passes through distinct phases of development during its individual life, like embryogenesis, growth and differentiation, followed by aging. The duration of each single phase is for any living cell predetermined by a biological clock that is presumably governed by genetic activity. While nobody doubts that embryogenesis and differentiation are not possible without genetic control, for the second half of an individual life, for many, this seems less indispensable. This appears, however, unjustified if we consider the strict regulation of the aging process that is evident in many respects.

Indeed, both the phase of development and the phase of aging have in common that they are strictly time dependent and irreversible. The difference can be seen in the fact that development is accompanied by an increase in fitness and the progression of a program, while aging is characterized by a loss of fitness and the regression of that same program.

In the meantime it has been elucidated that the clock that determines aging is not somewhere in the organism but in each individual cell.

While for a long time it was believed that isolated cells in culture were immortal, it has been shown by Hayflick that this is not true (9). Human fibroblasts undergo a limited number of cell divisions in cell culture and then die from inner causes. Immortality instead is confined to transformed cells which have escaped normal control, such as cancer cells.

It therefore might turn out that the genetic control of the aging process is only a special case of the overall problem of gene regulation and that a full understanding of aging will have to await a more complete insight into these fundamental, but still mysterious, biological connections.

References

1. Hufeland, CW (1798) Die Kunst das menschliche Leben zu verlängern. Akademische Buchhandlung, Jena
2. Weismann, A (1884) Über Leben und Tod. Jena
3. Goethe, JW von (1783) Fragment „Die Natur“, Tiefurter Journal
4. Wood, WB (1988) The nematode *Caenorhabditis elegans*. Cold Spring Harbor Laboratory , Cold Spring Harbor
5. Reichel, W et al (1971) Progeria and Werner's syndrome as models for the study of normal human aging. J Am Geriatric Soc 19: 369 – 375
5. Dayhoff, MO (1972) Atlas of Protein Sequence and Structure Vol. 5, National Biomedical Res. Found., Washington, D.C.
7. Orgel, LE (1970) The maintenance of the accuracy of protein synthesis and its relevance to ageing: a correlation. Proc Natl Acad Sci USA 67: 1476
8. Curtis, HJ (1968) Das Altern – die biologischen Vorgänge. Gustav Fischer, Stuttgart
9. Hayflick, L (1980) The cell biology of human aging. Sci Am 242: 58 – 65

Enzymatic and DNA/Protein Interaction Studies Indicate That Uracil in Neuronal DNA Could Contribute to Nerve Cell Aging*

F. Focher, P. Mazzarello, A. Verri, G. Biamonti, and S. Spadari

Introduction

Several hypotheses have been proposed to explain the molecular mechanisms of aging. Common to some of these hypotheses is that aging is the result of an accumulation of stochastic damage to DNA [16]. Damaged DNA is continuously repaired by several mechanisms in both proliferating and resting cells in order to avoid defective replication and abnormalities in transcription and translation [7]. One sort of such damage is the presence of uracil in DNA, deriving either from spontaneous or induced deamination of cytosine to uracil [7] or from uracil misincorporation during DNA replication in place of thymine (Fig. 1), as has been described in some viruses, bacteria, and cultured human lymphocytes.

It was also observed in lymphocytes that increased levels of dUTP, due to block of thymidylate synthetase by methotrexate, favor the misincorporation of uracil into DNA [9, 10].

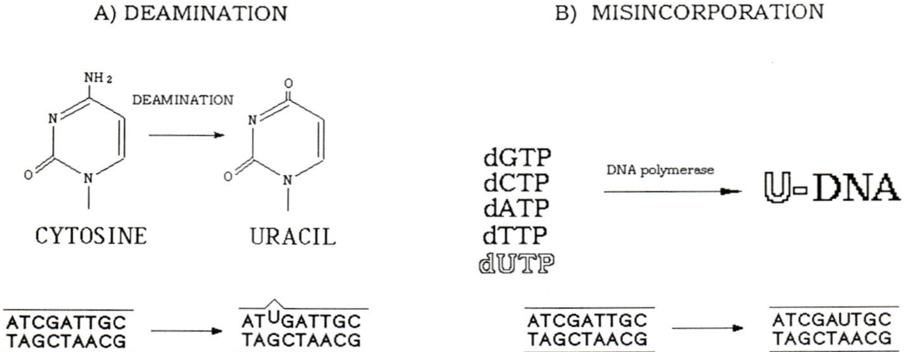

Fig. 1. Uracil in DNA arises either from cytosine deamination (leading to a mismatch) or from dUTP misincorporation in place of dTTP during DNA synthesis (leading to a correct base pairing)

* The CNR Progetto Finalizzato Biotecnologie e Biostrumentazione and the Associazione Italiana Sclerosi Laterale Amiotrofica (B. Vieri grant) provided fellowships to A. V. and P. M., respectively. This work was supported by the C.N.R. Progetti Finalizzati Biotecnologie and Chimica Fine.

Reihe der Villa Vigoni, Band 1
Biology of Aging
Zwilling and Balduini (Eds.)

When present in DNA, uracil is rapidly removed by the action of uracil-DNA glycosylase, which cleaves the N-glycosylic bond linking the uracil to the deoxyribose phosphate backbone. This enzyme is highly represented in proliferating mammalian cells and is cell-cycle regulated, showing an increase during S phase [11, 25].

Recently [5], we investigated: (1) the ability of DNA polymerase β to incorporate dUTP into DNA; (2) the perinatal levels of nuclear uracil-DNA glycosylase and of two enzymes that modulate the pool size of dUTP, namely dUTPase and nucleoside diphosphokinase (NDP kinase); and (3) the effect of the presence of uracil in DNA on DNA/protein interaction, using as a model the cAMP responsive element (CRE) sequence [29].

DNA Polymerase β Could Incorporate dUTP in Neuronal DNA

The misincorporation in DNA of uracil in place of thymine depends upon the action of DNA polymerases which can use dUTP as substrate instead of dTTP. Mammalian cells possess at least five different DNA polymerases named α, β, γ, δ and ϵ [1]. DNA polymerases α, δ and ϵ appear essential for nuclear DNA replication [1, 4, 5, 8, 26, 27], DNA polymerase β is mainly associated with DNA repair [4, 30], and DNA polymerase γ is the mitochondrial DNA replicating enzyme [12–14]. Some years ago, we demonstrated that adult neurons lack DNA polymerase α [14] and, more recently, that they also lack DNA polymerase δ/ϵ [26] (Fig. 2), and this correlates well with the fact that at birth neurons stop DNA replication, they cease to divide, and once destroyed either by aging or disease, they cannot be replaced. Figure 2 also shows that DNA polymerase β and γ are constant during neuronal development. DNA polymerase γ is confined to mitochondria, and adult neuronal

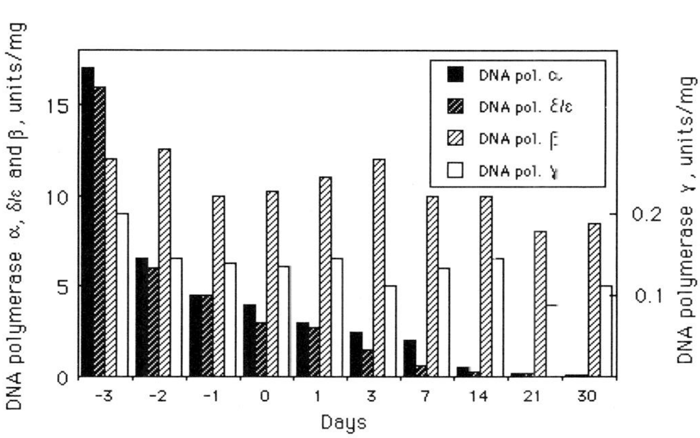

Fig. 2. Activity profiles of DNA polymerases α, β, γ, and δ during nerve cell development. Our DNA polymerase ϵ assay, under these conditions, allows measurement of both DNA polymerase δ and ϵ

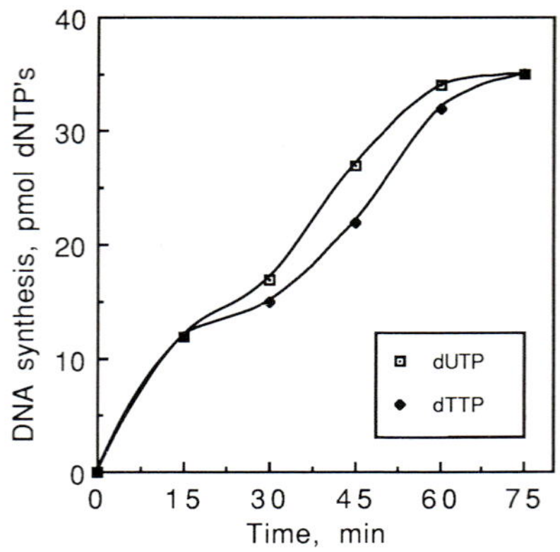

Fig. 3. In vitro parallel DNA synthesis kinetics with DNA polymerase β indicate that dTTP and dUTP are incorporated with the same efficiency

nuclei contain exclusively DNA polymerase β [14, 30]; this enzyme can repair damaged nuclear DNA [30]. Among the known eukaryotic DNA synthesizing enzymes, DNA polymerase β appears in vitro the most error prone in the selection of the complementary base during DNA synthesis [17, 18, 28]. Compared with calf thymus DNA polymerase α or *E. coli* polymerase I, which have an error rate of approximately 1/500 000 – 1/700 000, DNA polymerase β has a very low accuracy, showing an error rate of 1/8 000. Thus, we tested the capability of purified rat neuronal DNA polymerase β to incorporate dUTP in place of dTTP into DNA during in vitro DNA synthesis in the presence of the incorrect substrate dUTP. Figure 3 shows that DNA polymerase β can incorporate dTTP as well as dUTP and with the same efficiency.

We evaluated the K_m for dUTP and dTTP of rat neuronal DNA polymerase β and compared them with those obtained with other DNA polymerases, namely human DNA polymerases α, β, and ϵ. The results of these studies indicate that DNA polymerase β is unable to discriminate between the two substrates, and therefore the incorporation of dUTP into DNA can occur in adult neurons, depending on the relative concentration of dUTP. In fact, DNA polymerase β either from rat neurons or from HeLa cells has an identical K_m for dTTP and dUTP (2.5 μM for the rat and 2 μM for the human enzyme, respectively), whereas DNA polymerase α and DNA polymerase ϵ are more discriminating enzymes, showing a different K_m for these two substrates (11 μM dTTP and 29 μM dUTP for DNA polymerase α and 1 μM dTTP and 2 μM dUTP for DNA polymerase ϵ [6]).

Developmental Activity Profiles of dUTPase, NDP kinase, and uracil-DNA glycosylase in Neurons

It is reasonable to hypothesize that DNA in adult neurons could contain uracil arising either from spontaneous cytosine deamination or, as we have just seen, from misincorporation during DNA repair synthesis due to DNA polymerase β. The consequent mutagenic effects of these phenomena could be minimized by (1) low intracellular levels of dUTP and by (2) the action of uracil-DNA glycosylase, the enzyme that catalyzes the specific removal of uracil from DNA. Therefore, we studied the developmental activity profiles of uracil-DNA glycosylase, dUTPase, and NDP kinase, enzymes that control the pool size of dUTP.

The enzymatic activities of these three enzymes were determined in neurons from day 3 before birth up to day 30 after birth. The results, presented in Fig. 4, indicate that the activity profile of dUTPase slightly increases during neuronal development, whereas that of NDP kinase slightly decreases. This demonstrates the relative stability and, probably, the low level of dUTP in neurons. More interesting is the observation that the enzyme that directly removes the uracil from DNA, uracil-DNA glycosylase, shows a drastic age-dependent decrease of its activity levels; the pattern very much mimics that of the replicative DNA polymerases α and δ/ϵ [14, 26]. This is consistent with the observation by Seal and Sirover [24], who found a physical association of uracil-DNA glycosylase with DNA polymerase α.

The possibility that unspecific nucleases present in the nuclear extract could mimic the uracil-DNA glycosylase activity in our assay has been ruled out by testing in parallel the neuronal extracts on a DNA template labelled with [^{3}H]thymidine and by analyzing the products of the reaction on TLC-PEI chromatography. The resistance of the [^{3}H]thymidine-labelled DNA to the action of nuclear extracts and the fact that the radioactivity removed from [^{3}H]uracil-labelled DNA was, as expected, fully associated with uracil and not with dUMP indicate that our assay specifically tests the uracil-DNA glycosylase activity [6].

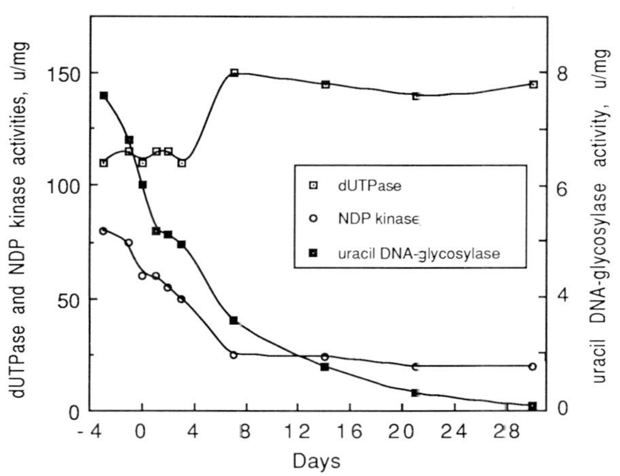

Fig. 4. Activity profiles of dUTPase, nucleoside diphosphokinase (NDP) kinase, and uracil-DNA glycosylase during nerve cell development

```
                   -4 -3 -2 -1  1  2  3  4          upper
5'  GATCCTTGGC | T  G  A  C  G  T  C  A | GAGAGAGA   3'
3'    AGGAACCG | A  C  T  G  C  A  G  T | CTCTCTCT   5'
                  4  3  2  1 -1 -2 -3 -4            lower
```

Fig. 5. Synthetic oligonucleotide corresponding to the cAMP responsive element (CRE) in rat somatostatin gene

Effect of the Presence of Uracil in DNA on DNA/Protein Interactions

Our studies allow us to hypothesize that in adult neurons the possible misincorporation of uracil by DNA polymerase β or the spontaneous deamination of cytosine to uracil coupled with the lack of uracil-DNA glycosylase could lead to a progressive increase in the level of uracil in the genome during a lifespan [19].

We propose here that the age-related increasing levels of uracil in DNA could be involved in the biological process of aging and neurodegeneration. The presence of uracil, in fact, could affect gene expression regulation by altering the sequence-specific binding of transcription factors. This is suggested by studies on the effect of the substitution of cytosine and thymine with uracil on protein/DNA interactions using, as a model, the CRE of the rat somatostatin gene promoter (Fig. 5). The CRE sequence is an inducible enhancer/promoter element of genes that induces transcription in response to an increase in the intracellular cAMP concentration [23]. CRE is also present in promoters of genes specific to neuronal cells, such as those encoding preproencephalin [2] and somatostatin [21].

The double-stranded synthetic oligonucleotide shown in Fig. 5, corresponding to a fragment of the rat somatostatin gene containing the CRE sequence (TGACGTCA), and oligonucleotides in which point mutations were introduced in position -4 (T→U) and -1 (C→U), in order to mimic uracil deriving either from misincorporation or cytosine deamination, respectively, were challenged with a HeLa cell nuclear extract prepared according to Dignam et al. [3] in band shift assays [29].

Figure 6 indicates that T→U substitution, deriving from uracil misincorporation in position -4 in both strands of CRE, reduces the binding affinity, suggesting that the methyl group of T is essential for the specific recognition of the sequence.

The deamination of cytosine in the central position CpG of the CRE sequence also gave interesting results (Fig. 6). In fact, we have observed that the change to uracil, when it occurs either in the upper (oligo 52) or in lower strand (oligo 16), increases about threefold the binding affinity for CRE, which is further enhanced to tenfold when the deamination is present in both strands (oligo 56) (Figs. 6, 7).

The specific binding to oligo 56, shown in Fig. 7, is hardly challenged by oligo 12 (wild-type CRE), whereas competes with oligo 12 a low excess of oligo 56. This suggests that the binding to oligo 56 is probably due to the same

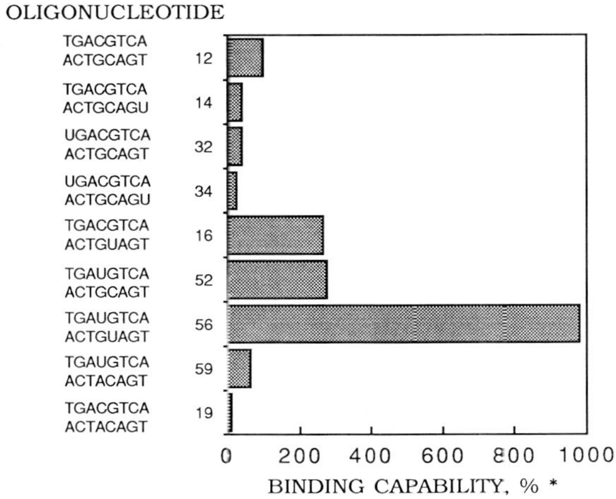

Fig. 6. Effect of uracil misincorporation and cytosine deamination on cAMP responsive element (CRE) binding factor(s)/CRE binding affinity. *Results were normalized to 100 % for binding to unmodified (wt) CRE sequence (oligo 12)

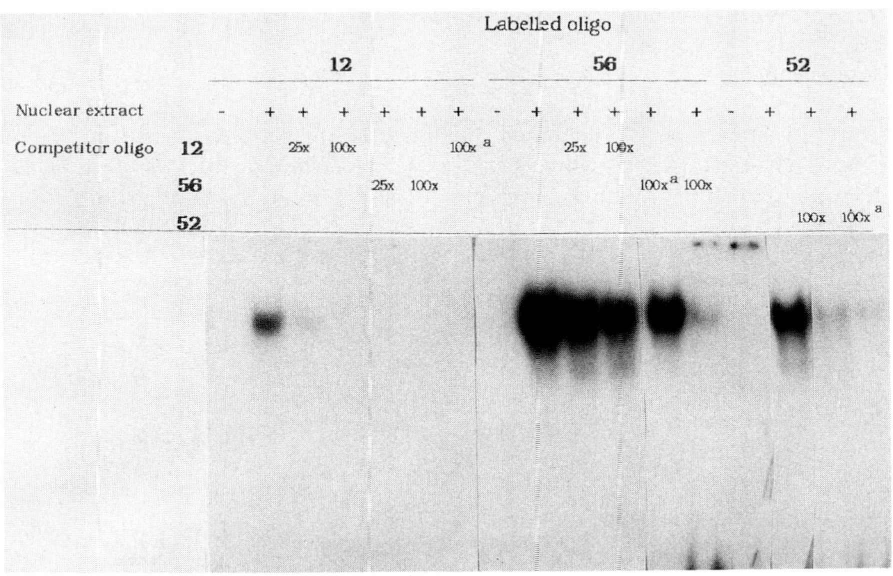

Fig. 7. Competition assays between oligo 12 (wild-type cAMP responsive element, CRE) and oligo 56 and 52. [a]A100-fold excess of unlabelled oligo was added to the mixture (nuclear extract/labelled oligo) 20 min after the start of reaction and the reaction was then continued for a further 20 min [29]

protein and that it is very strong, if not irreversible. Consistent with these data were the results obtained from the experiments shown in the same figure, in which 20 fmole of labelled oligo 56 were diluted 100-fold with unlabelled oligo 56, either immediately or after 20 min of incubation with the nuclear extract. The unlabelled oligo 56, when subsequently added, hardly competed with the bound labelled one. In parallel, experiments using oligo 12 and oligo 52 were performed: in these cases, we observed that labelled oligos 12 and 52, added immediately or after 20 min of incubation, were competetively challenged easily by the excess of unlabelled counterparts. The increased binding to oligos 16, 52, and mainly 56 could be due either to the mismatch alone or to the presence of U in position −1. In order to rule out this possibility, we studied oligos 59 and 19, in which an U without mismatch and a mismatch without U, respectively, are present. In both cases, the point mutations do not increase the affinity of the protein(s) for the sequence (see Fig. 6), and this suggests that the protein recognizes the G in position 1 and, once bound, is blocked by the mismatch in position −1 due to C→U conversion. Along the same line of evidence are the data of Iguchi-Ariga and Schaffner [15], who demonstrated that the CpG central dinucleotide of CRE has a crucial role in the correct functional recognition of the sequence.

Our findings showing that the substitutions T→U in position −4, C→U in position −1, and G→A in position 1 do not differ in effect when they occur in the upper or in the lower strand indicate that there is a symmetrical recognition of CRE by the CRE binding factor(s).

Conclusions

Neuronal aging may be related to the abnormal presence of uracil in DNA. This hypothesis is based on the following evidence (Fig. 8): (1) cytosine in DNA spontaneously deaminates to uracil [7]; (2) the only nuclear DNA poly-

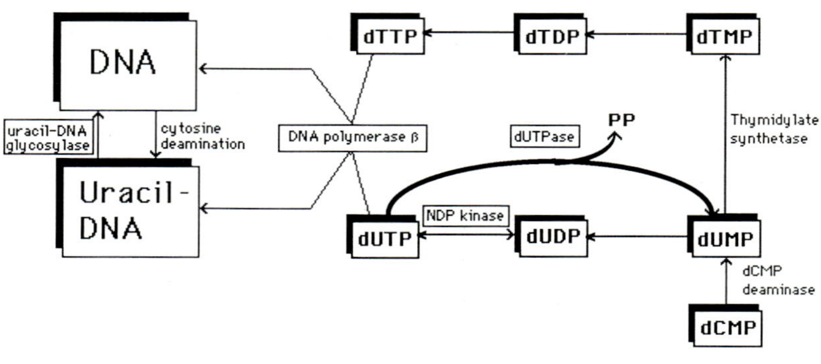

Fig. 8. Pathways leading to the presence of uracil in neuronal DNA. In adult neurons the lack of uracil-DNA glycosylase could impair the removal of uracil from DNA. *NDP*, nucleoside diphosphokinase

merase present in adult neurons, DNA polymerase β, which is able to repair damaged DNA, cannot discriminate between dTTP and dUTP, having the same K_m for both substrates (thus the incorporation of dUTP into neuronal DNA could be solely dependent on its relative intracellular concentration) [6]; (3) at birth, when neurons stop proliferating, uracil-DNA glycosylase, the enzyme responsible for the removal of uracil from DNA, nearly disappears [5]; (4) uracil in DNA, deriving either from misincorporation or from cytosine deamination, dramatically alters the specific recognition of DNA sequences by enzymes and/or by regulatory DNA binding proteins [29].

What we have observed for the CRE could be a general phenomenon which can occur in other regulatory sequences and early in the cascade of events leading to aging.

References

1. Burgers PMJ, Bambara RA, Campbell JL, Chang LMS, Downey, KM, Hübscher U, Lee MYWT, Linn SM, Spadari S (1990) Revised nomenclature for eukaryotic DNA polymerases. Eur J Biochem 191: 617 – 618
2. Comb M, Birnberg NC, Seasholtz A, Herbert E, Goodman HM (1986) A cyclic AMP- and phorbol ester-inducible DNA element. Nature 32: 353 – 356
3. Dignam JD, Lebovitz RM, Roeder RG (1985) Accurate transcription initiation by RNA polymerase II in a soluble extract from isolated mammalian nuclei. Nucleic Acids Res 11: 1475 – 1489
4. Focher F, Ferrari E, Spadari S, Hübscher U (1988) Do DNA polymerases δ and α act coordinately as leading and lagging strand replicases? FEBS Lett 229: 6 – 10
5. Focher F, Gassmann M, Hafkemeyer P, Ferrari E, Spadari S, Hübscher U (1989) Calf thymus DNA polymerase δ independent of proliferating cell nuclear antigen (PCNA), Nucleic Acids Res 17: 1805 – 1821
6. Focher F, Mazzarello P, Verri A, Hübscher U, Spadari S (1990) Activity profiles of enzymes that control the uracil incorporation into DNA during neuronal development. Mutat Res 237: 65 – 73
7. Friedberg EC (1985) DNA repair. Freeman, New York
8. Fry M, Loeb LA (1986) Animal cell DNA polymerases, CRC, Boca Raton
9. Goulian M, Bleile B, Tseng BY (1980) Methotrexate-induced misincorporation of uracil into DNA. Proc Natl Acad Sci USA 77: 1956 – 1960
10. Goulian M, Bleile B, Tseng BY (1980) The effect of methotrexate on levels of dUTP in animal cells. J Biol Chem 255: 10630 – 10637
11. Gupta PK, Sirover MA (1981) Stimulation of the nuclear uracil DNA glycosylase in proliferating human fibroblasts. Cancer Res 41: 3133 – 3136
12. Hübscher U, Kuenzle CC and Spadari S (1977) Variation of DNA polymerases α, β and γ during perinatal tissue growth and differentiation. Nucleic Acids Res 4: 2917 – 2929
13. Hübscher U, Kuenzle CC, Limacher W, Scherrer P, Spadari S (1978) Functions of DNA polymerase α, β and γ in neurons during development, Cold Spring Harbor Symp Quant Biol 43: 625 – 629
14. Hübscher U, Kuenzle CC, Spadari S (1979) Functional roles of DNA polymerase β and γ. Proc Natl Acad Sci USA 76: 2316 – 2320
15. Iguchi-Ariga, SMM, Schaffner W (1989) CpG methylation of the cAMP-responsive enhancer/promoter sequence TGACGTCA abolishes specific factor binding as well as transcriptional activation. Genes Dev 3: 612 – 619
16. Kirkwood TBL (1989) DNA, mutation and aging, Mutat Res 219: 1 – 7

17. Kunkel TA, Loeb LA (1980) On the fidelity of DNA replication: the accuracy of *Escherichia coli* DNA polymerase I in copying natural DNA in vitro. J Biol Chem 255: 9961 – 9966
18. Kunkel TA, Loeb LA (1981) Fidelity of mammalian DNA polymerases. Science 213: 765 – 767
19. Mazzarello P, Focher F, Verri A, Spadari S (1990) Misincorporation of uracil into DNA as possible contributor to neuronal aging and abiotrophy. Int J Neurosci 50: 169 – 174
20. Medvedev ZA (1990) An attempt at a rational classification of theories of ageing. Biol Rev 65: 375 – 398
21. Montminy MR, Sevarino KA, Wagner JA, Mandel G, Goodman RH (1986) Identification of a cyclic-AMP-responsive element within the rat somatostatin gene. Proc Natl Acad Sci USA 83: 6682 – 6686
22. Mullaart E, Lohmann PHM, Berends F, Vijg J (1990) DNA damage metabolism and aging. Mutat Res 237: 189 – 210
23. Roesler WJ, Vandenbark GR, Hanson RW (1988) Cyclic AMP and induction of eukaryotic gene transcription. J Biol Chem 263: 9063 – 9066
24. Seal G, Sirover MA (1986) Physical association of the human base-excision repair enzyme uracil DNA glycosylase with the 70000 dalton catalytic subunit of DNA polymerase α. Proc Natl Acad Sci USA 83: 7608 – 7612
25. Sirover MA (1979) Induction of the DNA repair enzyme uracil DNA-glycosylase in stimulated human lymphocytes, Cancer Res 39: 2090 – 2095
26. Spadari S, Focher F, Hübscher U (1988) Developmental activity profile of DNA polymerases δ and α in rat neurons suggests a coordinated in vivo function. In Vivo 2: 317 – 320
27. Spadari S, Montecucco A, Pedrali-Noy G, Ciarrocchi G, Focher F, Hübscher U (1989) A double-loop model for the replication of eukaryotic DNA. Mutat Res 219: 147 – 156
28. Subba Rao K, Martin GM, Loeb LA (1985) Fidelity od DNA polymerase-β in neurons from young and very aged mice. J Neurochem 45: 1273 – 1278
29. Verri A, Mazzarello P, Biamonti G, Spadari S, Focher F (1990) The specific binding to the cAMP responsive element (CRE) sequence (TGACGTCA) is reduced by the misincorporation of U and increased by the deamination of C. Nucleic Acids Res 18: 5775 – 5780
30. Waser J, Hübscher U, Kuenzle CC, Spadari S (1979) DNA polymerase ß from brain neurons is a repair enzyme. Eur J Biochem 97: 361 – 368

Age-Associated Changes of Muscarinic Cholinergic and *N*-Methyl-D-Aspartate Receptors in the Mouse Brain: Partial Reconstitution by Phosphatidylserine Treatment

W.E. Müller, C.M. Gelbmann, S.A. Cohen, L. Stoll, T. Schubert, and H. Hartmann

Receptor Deficits and the Membrane Hypothesis of Brain Aging

The so-called membrane hypothesis of aging assumes that several functional deficits of the aging brain can be explained on the basis of age-related changes of the composition of brain cell membranes leading to an increased cholesterol to phospholipid ratio, increased levels of saturated fatty acids, and a decreased membrane fluidity [3, 22]. Since the integrity of the neuronal membrane seems to be fundamental for most biochemical mechanisms involved in cell to cell communication like transmitter release, receptor activation, and signal transduction, many functional deficits of the aging brain can be easily explained by this hypothesis [3].

One of the most important deficits of the aged human or animal brain is impairment of memory and other cognitive functions. The biochemical basis of this is not yet finally understood, but an involvement of cholinergic and glutamatergic (via *N*-methyl-D-aspartate (NMDA) receptors) mechanisms might be possible, because of their important role in cognition [7, 15]. Accordingly, our current working hypothesis assumes that age-related changes of the neuronal membrane leading to dysfunctions of muscarinic and glutamatergic (NMDA-type) receptors are one of the major biochemical abnormalities responsible for impaired cognitive performance in aged animals and humans. Furthermore, drugs which correct age-associated membrane deficits should also be beneficial for treating age-associated memory impairment.

Certainly the most thoroughly investigated drug in this respect is the compound phosphatidylserine (BC-PS) which represents a natural membrane constituent of central neurons [3, 4]. Chronic BC-PS treatment of aged animals has been demonstrated to normalize a variety of biochemical deficits of the aging brain [4]. Additionally, improvements of memory and other aspects of cognitive performance in animals [23, 26] and in demented patients [1] have been observed. As a possible explanation for these beneficial effects in aged animals, a normalization of age-related decrease of membrane fluidity due to an increase of the cholesterol to phospholipid ratio of brain cell membranes has been reported [3], but the final relationship between the effects of BC-PS at the level of the composition of the neuronal membrane and its effects on cognitive deficits is not yet well understood. Our working hypothesis assumes

Reihe der Villa Vigoni, Band 1
Biology of Aging
Zwilling and Balduini (Eds.)

that BC-PS improves cognitive functions by ameliorating membrane properties which in turn restore deficits of central muscarinic cholinergic and NMDA receptors.

Age-Associated Loss of Neurotransmitter Receptors in the Mouse Brain: Effects of Chronic Phosphatidylserine treatment

Muscarinic Cholinergic Receptors

In agreement with our previous report [19] and other findings in the mouse brain [10, 24], and similar to findings in human brain [16], muscarinic cholinergic receptor binding as determined by saturation experiments using the antagonist tritiated quinuclidinyl benzilate ([^{3}H]QNB) is decreased in brains of aged mice when compared with young mice (Table 1). The age-related deficit is most pronounced in the brainstem (− 31 %), intermediate in the forebrain (− 18 %), still present in the midbrain (− 12 %), but absent in the cerebellum (Table 1). The dissociation constants (K_D) of the radioligand are not different between young and aged animals, indicating no changes of receptor properties (Table 1). Similar age-related receptor deficits were found [12] when muscarinic cholinergic receptor density was determined using the quarternary antagonist tritiated *N*-methyl scopolamine ([^{3}H]NMS) (data not shown).

When young mice were treated chronically with BC-PS, muscarinic cholinergic receptor density was not altered (see Table 1). By contrast, chronic treatment of aged animals with BC-PS under the same conditions increased specific [^{3}H]QNB binding by elevating receptor density (B_{max}) without affecting receptor affinity (K_D) (see Table 1). As indicated by ANOVA the effect of chronic treatment with BC-PS on muscarinic cholinergic receptor density was significant in all those brain areas in which age-related receptor deficits are present like forebrain, midbrain, and brainstem but not cerebellum, in which no age-related changes have been found (see Table 1). This observation is also supported by the post hoc analysis of the data using the *t*-test (see Table 1).

The effect of chronic treatment with BC-PS on muscarinic cholinergic receptor binding in the forebrain of aged mice is dependent on the dose of BC-PS (Fig. 1). A significant effect was already seen with 10 mg/kg (+ 15 %) which could be further augmented by increasing the dose to 20 mg/kg (+ 22 %) or to 40 mg/kg (+ 28 %). The effect was specific for BC-PS since chronic treatment with phosphatidylcholine even at the high dose of 40 mg/kg did not alter muscarinic cholinergic receptor density under the same experimental conditions (Fig. 1).

All effects of age or BC-PS treatment on the density of muscarinic cholinergic receptors in the mouse brain were similar whether density was expressed relative to milligram protein (see Table 1 and Fig. 1) or to the wet weight (data not shown).

Table 1. Effect of chronic treatment with phosphatidylserine (BC-PS) on specific tritiated quinuclidinyl benzilate ([^{3}H]QNB) binding in the brains of young and aged mice

Area	Young mice		Aged mice		Analysis of variance			
	K_D (nmol/l)	B_{max} (pmol/mg protein)	K_D (nmol/l)	B_{max} (pmol/mg protein)	Main effect	df	F	P
Forebrain								
Control	0.10 ± 0.02 (6)	$1.77 + 0.10$	0.09 ± 0.01 (6)	$1.46 \pm 0.10^{***}$	Age	1	48.60	0.0001
BC-PS	0.11 ± 0.02 (6)	1.80 ± 0.08	0.09 ± 0.01 (6)	$1.63 \pm 0.08^{++}$	Treatment	1	8.63	0.0081
					Age × treatment	1	3.72	0.0680
Midbrain								
Control	0.08 ± 0.01 (6)	1.21 ± 0.05	0.10 ± 0.01 (6)	$1.08 \pm 0.7^{*}$	Age	1	5.18	0.0340
BC-PS	0.08 ± 0.01 (6)	1.22 ± 0.04	0.10 ± 0.01 (6)	$1.24 \pm 0.08^{+++}$	Treatment	1	10.02	0.0049
					Age × treatment	1	9.31	0.0063
Brainstem								
Control	0.11 ± 0.01 (6)	0.48 ± 0.01	0.07 ± 0.02 (6)	$0.33 \pm 0.03^{***}$	Age	1	169.52	0.0001
BC-PS	0.09 ± 0.01 (6)	0.48 ± 0.02	0.09 ± 0.03 (6)	$0.37 \pm 0.02^{+}$	Treatment	1	3.00	0.0986
					Age × treatment	1	6.19	0.0218
Cerebellum								
Control	0.10 ± 0.01 (6)	0.13 ± 0.01	0.10 ± 0.02 (6)	0.13 ± 0.01	Age	1	2.47	0.1327
BC-PS	0.09 ± 0.02 (6)	0.13 ± 0.01	0.10 ± 0.02 (6)	0.13 ± 0.01	Treatment	1	0.02	0.8945
					Age × treatment	1	0.08	0.7757

Compared with the aged controls:
$^{+}$ $P < 0.05$
$^{++}$ $P < 0.01$
$^{+++}$ $P < 0.005$

Compared with the young controls:
* $P < 0.05$
*** $P < 0.001$

Note: The animals were treated with BC-PS (20 mg/kg, once daily i.p.) or saline (i.p.) for 3 weeks. Receptor density (K_D) and receptor affinity (B_{max}) of specific [^{3}H]QNB binding were determined as described elsewhere (12). Data are means $\pm$SD of (n) experiments, each representing an individual animal. Statistical comparison of B_{max} values was performed by two-way ANOVA and by t-test as post hoc analysis. Data are from Gelbmann and Müller [12]

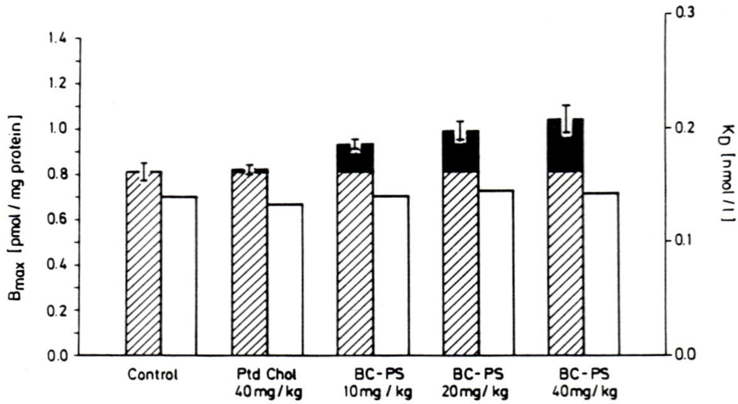

Fig. 1. Effect of chronic treatment with different doses of phosphatidylserine (BC-PS) on the density of specific tritiated *N*-methylscopolamine ([^{3}H]NMS) binding in the forebrain of aged mice. The animals were treated once daily i.p. for 3 weeks either with BC-PS at the doses indicated, or with phosphatidylcholine (ptd Chol; 40 mg/kg), or with saline. Data for K_D (*open columns*) or for B_{max} (*hatched columns*) were obtained from saturation experiments. Data are means ± SD of 5 or 6 experiments, each representing an individual animal. Intergroup differences (B_{max}) between controls and the three BC-PS groups were highly significant ($F = 25.98$; df = 19, $p < 0.001$, one-way ANOVA). The B_{max} values of all three BC-PS groups are significantly ($P < 0.001$) different from the controls; all other data are not. Data are from Gelbman and Müller [12]

NMDA Receptor

Similar to findings in the aged human cortex [13], the density of NMDA receptors in the forebrain of aged mice is decreased by about 35 % when compared with young animals (Fig. 2). While chronic treatment of young animals with BC-PS was without any effect on NMDA-receptor density (Fig. 2), the same treatment elevated it in the forebrain of aged animals by about 20 % (Fig. 3). One single treatment of aged animals with BC-PS was without any effect on NMDA-receptor density (Fig. 2).

Other Receptors

In order to prove the specificity of the effects of BC-PS treatment on age-related deficits of muscarinic cholinergic receptors and NMDA receptors (as parts of two neurotransmitter systems with important roles for cognitive functions), we investigated additionally the effects of chronic BC-PS treatment on the densities of some other receptor systems in the aged mouse brain.

We found a highly significant reduction of the density of benzodiazepine receptors in the midbrain of the aged NMRI mouse compared with young animals (Fig. 3). Chronic treatment with BC-PS (20 mg/kg for 21 days) had a

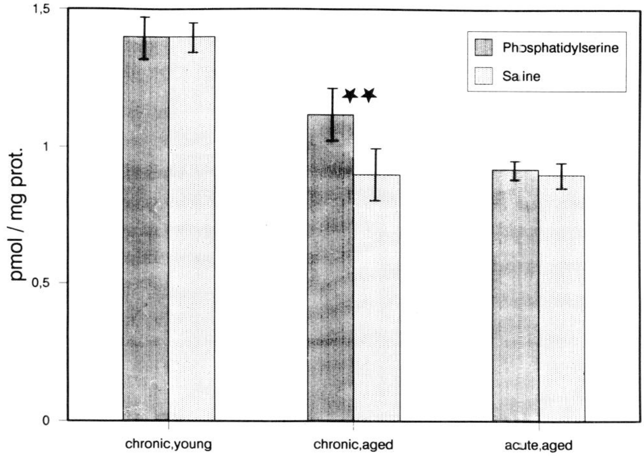

Fig. 2. Effects of chronic (21 days) or one single treatment on NMDA-receptor density in the forebrain of young (3 months) or aged (18 months) female NMRI mice. Data represent B_{max} as determined from saturation experiments using [^{3}H]MK 801 as radioligand. Data are means $\pm$SD from eight experiments. Data are from Cohen and Müller [5]. $^{**}p<0.01$

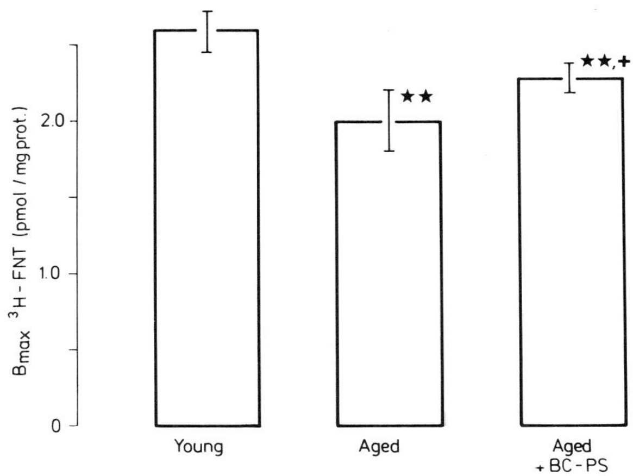

Fig. 3. Effects of aging and chronic treatment with phosphatidylserine (BC-PS; 20 mg/kg, 21 days) on the density of benzodiazepine receptors in the midbrain of the female NMRI mouse. B_{max} was determined from saturation experiments using [^{3}H]flunitrazepam ([^{3}H]FNT) as radioligand [9]. Data are means $\pm$SD of 6 experiments (Müller et al. unpublished data). ** $p<0.01$ when compared with young animals; $^+$ $p<0.05$ when compared with aged controls

small but significant effect on the benzodiazepine receptor density in aged mice with an elevation of B_{max} of about 10 % (Fig. 3). However, since a similar effect (about 10 % increase) has been reported for the chronic treatment of

Table 2. Effect of chronic phosphatidylserine (BS-PS) treatment on several receptor systems of the aged mouse brain (female NMRI mice). Data are taken from Gelbmann and Müller [12], Cohen and Müller [5], or from unpublished work of our laboratory

Receptor	Age-associated deficit	Effect of chronic BC-PS treatment (20 mg/kg for 21 days)
Muscarinic cholinergic	Decrease of B_{max} in the forebrain by 15 % – 25 %	Elevation of B_{max} by 15% – 25%
NMDA	Decrease of B_{max} in the forebrain by 35%	Elevation of B_{max} by 20%
	Enhanced sensitivity to glycine and L-glutamate	Normalization of the sensitivity to glycine and L-glutamate
Benzodiazepine	Decrease of B_{max} in the midbrain by 20 %	Elevation of B_{max} by 10 %
α_2-Adrenoceptor	Decrease of high-affinity agonist sites by 30 % (whole brain) [11]	No effect
α_1-Adrenoceptor	No deficit (whole brain)	No effect

young rats with BC-PS [14], the effect of BC-PS on benzodiazepine receptor density in aged mice might not be specific for age-related deficits, as has been demonstrated for the muscarinic cholinergic receptor and the NMDA receptor.

Chronic treatment of aged mice with BC-PS was without any effect on the densities of α_1-and α_2-adrenergic receptors in the mouse brain (Table 2). It is noteworthy that for these two systems age-related deficits of receptor densities could not be demonstrated.

Conclusion

The findings summarized in the present communication about the effects of BC-PS on several receptor deficits of the aged mouse brain support many other observations indicating that a variety of age-related deficits of brain function might specifically benefit from chronic treatment with BC-PS [3, 4, 17, 18, 26]. The effects of chronic BC-PS treatment on the densities of muscarinic cholinergic and NMDA receptors seem to be specific for the age-associated decline in receptor number, since it is not present in young animals and can be observed in older animals only in those brain regions in which age-associated deficits of receptor density are present. The specificity of the effect is further demonstrated by the failure to influence receptor density by chronic treatment with phosphatidylcholine in older animals and by the failure to alter the densities of α_1- or α_2-adrenoceptors by BC-PS treatment, two systems for which no age-associated deficits of receptor number are present in the aged mouse brain. The effects of chronic BC-PS treatment on the densities of muscarinic cholinergic and NMDA receptors can be observed under treatment conditions (dosage and time) which have been repetitively demonstrated to improve memory and other cognitive functions in aged mice and rats [8, 23, 26]. On the other hand, the effects might not be restricted to these two systems, since other age-related receptor deficits might also benefit from BC-PS treatment. This is supported by recent findings in the rabbit brain, where chronic BC-PS treatment has been demonstrated to elevate the density of prolactin receptors only in those regions in which age-related deficits of receptor density have been found [6].

Since a large variety of different observations strongly indicate that cholinergic and NMDA receptor-mediated glutamatergic mechanisms are of major importance for memory and other cognitive functions [7, 15], it is certainly very attractive to speculate that the elevation of receptor densities reported in the present communication is involved per se in the positive effects of chronic BC-PS treatment on cognitive performance of older animals. However, one has to be very careful in interpreting variations of receptor density in terms of changes of receptor function, especially if the alterations of density are as small as those found in our experiments. Accordingly, while the present data strongly support the general ability of BC-PS to restore age-related biochemical deficits of the CNS at the level of the neuronal membrane, the possible relevance of the findings as an explanation for the beneficial effects of the compound in cognitive deficits has not yet been proven by these data. However, this assumption would be strongly supported by findings of additional beneficial effects of BC-PS on the function of muscarinic cholinergic and NMDA receptors in the aging brain.

Phosphatidylserine and NMDA-Receptor Function in Aging

MK-801, the radioligand used by us to investigate the density of NMDA receptors in the brains of young and aged mice, does not interact directly with the L-glutamate recognition site of the NMDA receptor but binds to a specific site located inside the calcium channel, which is opened by NMDA-receptor activation [20, 25]. Accordingly, the access of MK-801 to this binding site and the affinity constant of its interaction with the binding site are strongly enhanced by the addition of L-glutamate to the incubation medium (Fig. 4 a). A very similar phenomenon can be observed by the addition of the amino acid glycine (Fig. 4 b). It is well-known from electrophysiological as well as receptor binding experiments that glycine interacts with an allosteric binding site of the NMDA receptor complex, which following this activation augments the response of NMDA-receptor agonists (opening of Ca channels; [20]). Accordingly, in binding experiments, glycine stimulates MK-801 binding in a fashion similar (Fig. 4 b) to the addition of L-glutamate (Fig. 4 a). Interestingly, both the L-glutamate effect as well as the glycine effect on MK-801 binding are

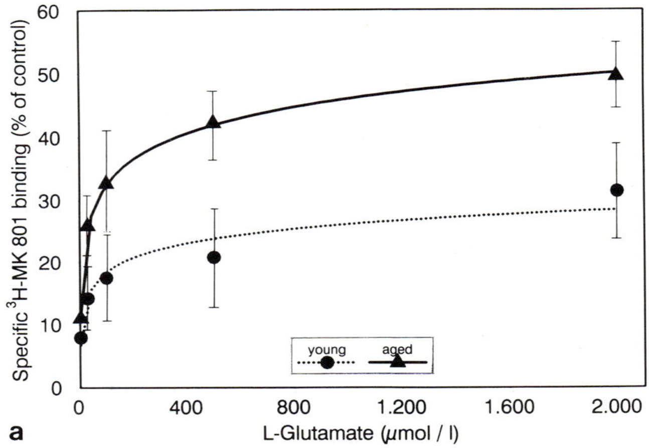

Fig. 4 a. Stimulation of specific [^{3}H]MK-801 binding to membranes of the forebrain of young and aged mice by increasing concentrations of L-glutamate. Data are means ± SD of 8 experiments. (From Cohen and Müller, [5]

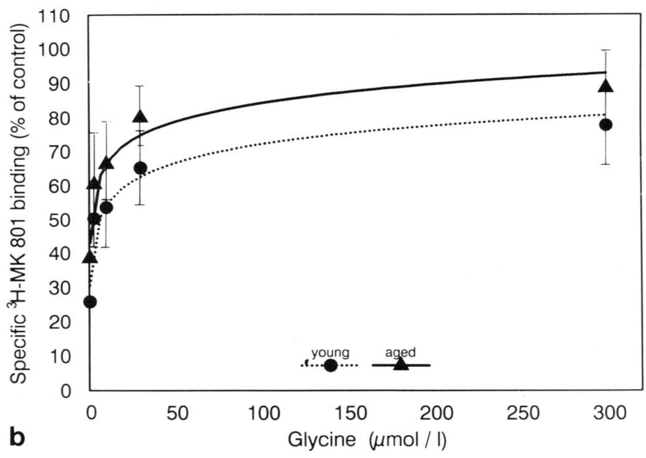

Fig. 4 b. Stimulation of specific [^{3}H]MK-801 binding to membranes of the forebrain of young and aged mice by increasing concentrations of glycine. Data are means ± SD of 8 experiments. (From Cohen and Müller [5])

much more pronounced in forebrain membranes of aged than of young mice (Fig. 4 a). The age difference is not only present for the maximal stimulation of MK-801 binding by high concentrations of both amino acids and by the concentrations needed to stimulate basal binding by 25 % (L-glutamate) or 50 % (glycine) but is also highly significant for the ED_{50} concentrations of

Table 3. Effects of age or BC-PS treatment (20 mg/kg, 21 days) on the stimulation of specific [^{3}H]MK-801 binding by L-glutamate and glycine. ED_{50} represents the concentration of both amino acids which yields 50 % maximal stimulation, and C_{25} or C_{50} are the concentrations which enhance baseline binding by 25 % or 50 %, respectively. (Data are from Cohen and Müller [5])

	L-Glutamate		Glycine	
	ED_{50} (μmol/l)	C_{25} (μmol/l)	ED_{50} (μmol/l)	C_{50} (μmol/l)
Young ($n=10$)	65 ± 33	728 ± 718	1.8 ± 1.3	11.9 ± 13
Aged ($n=10$)	17 ± 8**	26 ± 11**	0.6 ± 0.3*	1.8 ± 1.8*
Aged controls ($n=7$)	21 ± 9	34 ± 17	0.6 ± 0.3	0.5 ± 0.3
Aged + BC-PS ($n=7$)	61 ± 26**	764 ± 612**	1.6 ± 1.0*	2.0 ± 1.3*

* $p<0.05$, ** $p<0.01$ when compared with young or aged controls, respectively

both amino acids (concentrations needed to show 50 % maximal stimulation; Table 3).

While the enhanced maximal stimulation of [^{3}H]MK-801 binding by both amino acids might simply suggest that in the older mouse brain more calcium channels are in a closed state under baseline conditions, the reduced ED_{50} concentrations of both amino acids in the aged brain indicate enhanced affinities of both binding sites. At present, we interpret these findings as first evidence of NMDA receptor supersensitivity to both amino acids in aging, probably in order to compensate partially for the reduced density. Our experiments cannot be used to estimate whether or not NMDA-receptor-mediated transduction is still impaired in the aging brain or if the reduced density is sufficiently compensated by receptor supersensitivity. Recently published electrophysiological investigations in older rats seem to support the former possibility, assuming that NMDA-receptor function in the aging brain is reduced [2].

Chronic treatment of aged animals with BC-PS not only elevated NMDA-receptor density (see Fig. 2) but also reduced the supersensitivity of glycine and L-glutamate stimulation of MK-801 binding, so that the aged animals were no longer distinguishable from young ones (see Table 3). If we assume that supersensitivity of glycine and L-glutamate stimulation of MK-801 binding represents a compensatory mechanism for reduced NMDA-receptor function in older animals, the reversal of this effect by BC-PS suggests that the treatment normalizes NMDA-receptor function in aged animals. Final proof for this hypothesis might come from experiments on the effects of NMDA agonists on neuronal calcium signalling which are currently being performed in our laboratories.

Cholinergic Signal Transduction in the Aging Brain

As already mentioned, age-associated reductions of central muscarinic cholinergic receptor densities in the brains of animals and man are probably too small to explain fully the possible impairment of central cholinergic functions during normal aging. Certainly, the most attractive level to explain the age-associated deficits would be alterations of the responsiveness of these receptors. However, available data on muscarinic cholinergic receptor function in the aging brain are very conflicting, and our initial findings concerning a reduced response of phospholipase C to cholinergic stimulation in the aged mouse brain [21] have been supported by findings in aged Wistar rats but not in aged Fischer 344 rats (for review, see [16].

Because of these discrepancies we have recently started to evaluate the effect of aging on muscarinic cholinergic receptor function very carefully, using mechanically dissociated neurons of young (3 months) and aged (18 months) female NMRI mice. The experimental conditions have been improved compared with previous ones [21]. Cells are preincubated with [^{3}H]inositol for 1 h and are subsequently washed to remove free [^{3}H]inositol, in order to avoid possible interference by the additional incorporation of [^{3}H]inositol into membrane-bound phospholipids during stimulation by cholinergic agonists (Hartmann and Müller, unpublished data). Hopefully, these experiments will give us a final answer about the functional state of the cholinergic signal transduction in the aging brain.

Conclusions

The data reported clearly show that chronic BC-PS treatment specifically restores age-related deficits of several neurotransmitter receptors in the mouse brain. At present, we interpret these findings rather as further evidence that chronic treatment with this drug can normalize age-related alterations of membrane function than as a mechanism directly involved in the cognition-enhancing properties of BC-PS. On the other hand, due to the important role of NMDA receptors for cognition and because of our initial findings that BC-PS might enhance NMDA-receptor function in aged animals, the hypothesis that BC-PS improves memory and learning abilities by enhancing NMDA receptor function certainly merits further investigations.

Acknowledgement. The excellent technical assistance of Claudia Schäfer is gratefully acknowledged.

References

1. Amaducci L (1988) Phosphatidylserine in the treatment of Alzheimer's disease: results of a multicenter study. Psychopharmacol Bull 24: 130–134
2. Baskys A, Reynolds JN, Carlen PL (1990) NMDA depolarizations and long-term potentiation are reduced in the aged rat neocortex. Brain Res 530: 142–146
3. Calderini G, Aporti F, Bellini F, Bonetti AC, Rubini R, Teolato S, Xu C, Zanotti A, Toffano G (1985) Phospholipids as pharmacological tools in the aging brain. In: Horrocks LA (ed) Phospholipids in the nervous system, vol 2. Raven New York, pp 11–19
4. Calderini G, Bellini F, Bonetti AC, Galbiati E, Guidolin D, Milan F, Nunzi MG, Rubini R, Zanotti A, Toffano G (1987) Pharmacological properties of phosphatidylserine in the ageing brain. Clin Trials J 24: 9–17

5. Cohen SA, Müller WE (in preparation) Age-related alterations of NMDA-receptor properties in the mouse forebrain. Effects of chronic treatment with phosphatidylserine.
6. Di Carlo R, Muccioli G, Ghe C, Belussi G, Ravetti D, Genazzani E (1988) Phosphatidylserine activity on prolactin brain receptors of aged rabbits. Prog Neuropsychopharmacol Biol Psychiatry 12: 915 – 926
7. Drachman DA (1981) The cholinergic system, memory, and aging. In: Enna SJ (ed) Brain neurotransmitters and receptors in aging and age-related disorders. Raven, New York, pp 255 – 268
8. Drago F, Toffano G, Catalano L, Danielli R, Continella G, Scapagnini U (1983) Phosphatidylserine facilitates learning and memory processes in aged rats. In: Samuel D (ed) Aging of the brain. Raven, New York, pp 309 – 316
9. Fehske KJ, Zube I, Borbe HO, Wollert U, Müller WE (1982) β-Carboline binding indicates the presence of benzodiazepine receptor subclasses in the bovine central nervous system. Naunyn Schmiedebergs Arch Pharmacol 319: 172 – 177
10. Freund G (1980) Cholinergic receptor loss in brains of aging mice. Life Sci 26: 371 – 375
11. Gelbmann CM, Müller WE (1989) Specific decrease of highaffinity agonist states of alpha2-adrenoceptors in the aging mouse brain. J Neural Transm 79: 131 – 136
12. Gelbmann CM, Müller WE (in press) Chronic treatment with phosphatidylserine restores muscarinic cholinergic receptor deficits in the aged mouse brain. Neurobiol Aging
13. Kornhuber J, Mack-Burkhardt F, Konradi C, Fritze J, Riederer P (1989) Effect of antemortem and postmortem factors on ^{3}H-MK-801 binding in the human brain: transient elevation during early childhood. Life Sci 45: 745 – 749
14. Levi de Stein M, Medina JH, De Robertis E (1989) In vivo and in vitro modulation of central type benzodiazepine receptors by phosphatidylserine. Mol Brain Res 5: 9 – 15
15. Morris RGM (1989) Synaptic plasticity and learning: selective impairment of learning in rats and blockade of long-term potentiation in vivo by the *N*-methyl-D-aspartate receptor antagonist AP5. J Neurosci 9: 3040 – 3057
16. Müller WE, Stoll L, Schubert T, Gelbmann CM (1991) Central cholinergic functions and aging. Acta Psychiat Scand 83: Suppl. 366, 34 – 39
17. Nunzi MG, Milan F, Guidolin D, Toffano G (1987) Dendritic spine loss in hippocampus of aged rats. Effect of brain phosphatidylserine administration. Neurobiol Aging 8: 501 – 510
18. Pedata F, Giovannelli L, Spignoli G, Giovannini MG, Pepeu G (1985) Phosphatidylserine increases acetylcholine release from cortical slices in aged rats. Neurobiol Aging 6: 337 – 339
19. Pilch H, Müller WE (1988) Piracetam elevates muscarinic cholinergic receptor density in the frontal cortex of aged but not of young mice. Psychopharmacology 94: 74 – 78
20. Reynolds IJ, Murphy SN, Miller RJ (1987) 3H-labeled MK-801 binding to the excitatory amino acid receptor complex from rat brain is enhanced by glycine. Proc Natl Acad Sci USA 84: 7744 – 7748
21. Stoll L, Schubert T, Müller WE (in press) Age related deficits of central muscarinic cholinergic receptor function in the mouse brain. Partial restoration by chronic piracetam treatment. Neurobiol Aging
22. Sun AY, Sun GY (1979) Neurochemical aspects of the membrane hypothesis of aging. Interd Topics Gerontol 15: 34 – 53
23. Valzelli L, Kozak W, Zanotti A, Toffano G (1987) Activity of phosphatidylserine on memory retrieval and on exploration in mice. Meth Find Expl Clin Pharmacol 9: 657 – 660
24. Waller SB, London ED (1983) Age differences in choline acetyltransferase activities and muscarinic receptor binding in brain regions of C 57 BL/6J mice. Exp Gerontol 18: 419 – 425
25. Wong, EHF, Kemp JA, Priestley T, Knight AR, Woodruff GN, Iversen LL (1986) The anticonvulsant MK-801 is a potent *N*-methyl-D-aspartate antagonist. Proc Natl Acad Sci USA 83: 7104 – 7108
26. Zanotti A, Valzelli L, Toffano G (1989) Chronic phosphatidylserine treatment improves spatial memory and passive avoidance in aged rats. Psychopharmacology 99: 316 – 321

Membrane Processes in Human Red Cell Aging*

A. Brovelli, M.A. Castellana, G. Minetti, G. Piccinini, C. Seppi, M.R. DeRenzis, and C. Balduini

Introduction

During maturation, mammalian red cells lose organelles and subcellular structures; due to the absence of protein synthesis, circulating erythrocytes represent a good model for the study of how protein molecules and plasma membranes age. Human red cells spend 120 days in the circulation and are then removed in an age-dependent manner [1]. The presence of autoantibodies on the surface of senescent cells [2, 3] led to the development of the autoimmune hypothesis for senescent cell removal from the circulation and focus on plasma membrane studies about red cell aging. Since this presentation deals with membrane processes occurring during red cells senescence, the most important structural properties of plasma membrane in human red cells will be briefly summarized (Fig. 1). The most relevant ones concerning the integral domain of the membrane are the asymmetry in the distribution of phospholipids in the two halves of the bilayer and the topology and topography of integral proteins; in the peripheral domain, the supramolecular structure of the membrane skeleton and of its junctions with the integral domain are critical for the preservation of cell shape and deformability. It is interesting that the phospholipid distribution in the two halves of the bilayer, in particular phosphatidylserine, is entirely confined to the cytoplasmic side [5]; its exposure on the external surface increases cell adhesion to endothelium and macrophages [6]. Glycophorins and band 3 protein are the integral proteins most represented in quantitative terms (1×10^6 copies/cell). These glycoproteins under normal conditions are uniformly distributed on the cell surface, and this distribution is maintained and related to that of the peripheral proteins interacting with them [7]. As for their topology, glycophorin crosses the membrane once [8] and band 3, several times (the intralipid domain is the anion transport chan-

*Work supported by funds from MURST and CNR: Target Project 'Biotecnology'.

Note the abbrevations used are: *SDS-PAGE*, sodium dodecylsulfate-polyacrylamide-gel electrophoresis; *DACM*, N-(7-dimethyl-amino-4-methyl-coumarinyl)maleimide; *EDTA*, ethylenediaminetetraacetic acid, disodium salt; *MetSO*, methionine sulfoxide; *EM*, eosine-5-maleimide; *EITC*, eosine isothiocyanate; *DIDS*, 4,4'-diisothiocyanostilbene-2,2'-disulfonic acid disodium salt.

Reihe der Villa Vigoni, Band 1
Biology of Aging
Zwilling and Balduini (Eds.)

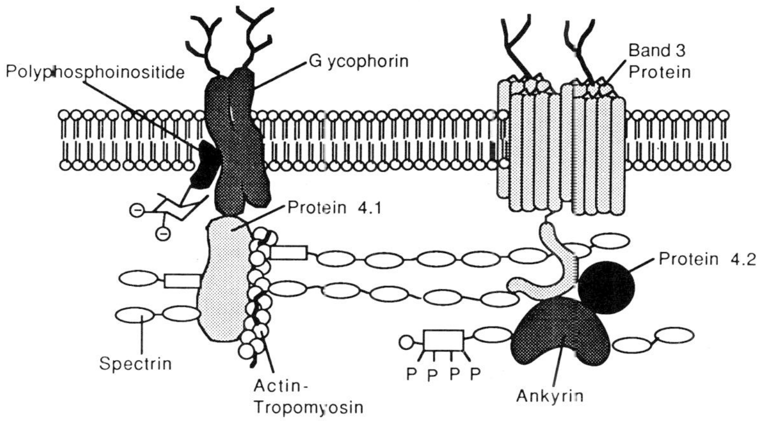

Fig. 1. Arrangement of integral and peripheral proteins in the human erythrocyte membrane (adapted from [4])

nel), and with its N-terminal cytoplasmic domain interacts with the membrane skeleton, hemoglobin, and some glycolytic enzymes [9].

Most peripheral proteins are organized in a net of filaments, the membrane skeleton, that maintains cell shape and deformability [10, 11]. The basic structural unit is represented by spectrin tetramers and oligomers interacting with 4.1 protein and actin protofilaments, and minor proteins (tropomyosin, adducin) stabilize this supramolecular structure [10, 11]. Junctions between the integral and peripheral domains are present at the level of band 3, that interacts with ankyrin which binds spectrin (band 4.2 protein stabilizes this interaction) [9, 12] and of glycophorin, that binds protein 4.1 via phosphatidylinositides [13]. Protein 4.1 contains a low affinity site for band 3 and a high affinity site for phosphatidylserine [14]. Structural damage to proteins of the junctions modifies cell shape and often produces membrane fragmentation and the release of vesicles [15].

The last event of red cell life seems to be the opsonization by autoantibodies triggering its phagocytosis [16]. Different signals have been recognized to be able to stimulate 'in vitro' phagocytosis of damaged or senescent cells [17]. Evidence is available that among these signals, epitopes on band 3 are specifically recognized by IgGs bound to the surface of senescent cells, and this new antigen has been called senescence antigen [19, 20]. On the other hand, Galili et al. [21, 22] have reported that the natural IgG antibody of normal plasma having a specificity for α-galactosyl residues binds to aged erythrocytes and that the antigens on the cell surface are carbohydrate chains of glycolipids [23, 24]. Recently, M. Clark [25] has confirmed that IgGs eluted from senescent cells recognize band 3, α-galactosyl residues and other epitopes still to be defined, thus reevaluating the role of epitopes different from those exposed on the band 3 molecule in determining antigenicity of the plasma membrane in senescent human red cells. Although the involvement of band 3-specific antibodies in the recognition of old red cell was suggested by many authors, their view diverges with regard to the mechanism leading to the exposure of the senescence antigen. Some authors presented evidence that this antigen is produced by proteolysis of band 3 [18, 19, 26]; others suggested that oxidation leads to band 3 oligomerization and to the exposure of senescence antigen [20, 27 – 29]. As for the exposure of galactosyl residues on senescent red cell membrane, some authors have presented proof of exposure of α- [21, 22] and others of β- [30 – 32] galactosyl sites.

The development of the autoimmune hypothesis for senescent cell removal from the circulation raised questions about the presence of senescence markers on the cell surface that permit such recognition and about the mechanisms of their development during the red cell life span. Studies on surface changes occurring during red cell senescence have been carried out mainly on density-separated red cells, since the densest red cell subpopulation is enriched in senescent cells [16]. Cellular processes involving the membrane and occurring in the dense cell population are reduction in membrane surface area [33 – 35], decrease of cell deformability [36 – 39], and a local modification of phospholipid asymmetry [6]. At the level of the membrane, slight modifications of the covalent structure of components have been described, produced by processes like oxidation [40 – 42], proteolysis [26, 43], glycation [44], methylation and transamidation [45], and modification of topology and topography of proteins have been reported or hypothesized [20, 27, 28, 46]. Most of these modifications are effective in promoting autoantibody binding and/or phagocytosis in vitro, thus supporting a possible role for each of these mechanisms in determining recognition and removal of senescent cells. Investigations carried out with in vivo [47, 48] and in vitro [49, 50] models of red cell senescence have shown that oxidation is one of the likely candidates for the mechanisms of senescence antigen formation, and characterization of density-separated cells has provided evidence that denser cells have suffered cumulative oxidative damage [16, 42, 51]. Although clues of oxidative processes are evident on the membrane of the senescent cell, the oxidative state of membrane proteins in human red cells of different ages was not investigated in detail in the past. Therefore, we have tried to quantitate the oxidative lesion the membrane proteins undergo during a red cell life span, in an attempt to understand what kind of processes expressed in the senescent red cell can be related to oxidation.

Diamide treatment of dog [47] and human [44] red cells has been used as a model of cell senescence. Diamide is an oxidant of thiol groups [52], and treatment of human red cells with micromolar diamide concentrations was shown to induce, like in senescent cells, the binding to the cell surface of band 3-specific IgGs and the complement-dependent phagocytosis of these cells [49]. Since diamide treatment of red cells has been used to mimic aging at the membrane level, we studied oxidative processes concerning membrane proteins and occurring in diamide-treated red cells and compared them with the oxidative processes evident in senescent red cell membrane.

Results

Oxidation of Membrane Proteins During Red Cell Life Span

Oxidative lesion to membrane proteins was studied in human erythrocytes of different ages, density separated by centrifugation in autologous plasma [53 – 55] and characterized for their creatine content as a cell-age parameter [56, 57], and was evaluated in ghost membranes prepared from young, middle-aged, and senescent cells by assaying their thiol [58] and methionine sulfoxide (MetSO) content [54] and in intact cells [58] by measuring the amount of the thiol-reagent *N*-(7-dimethylamino-4-methylcoumarinyl)maleimide (DACM) bound in situ by each membrane protein. Results obtained show that during aging of normal cells the oxidative state of membrane proteins increases (Fig. 2). This was better shown by the assay of MetSO residues than by thiol titration, when studies were carried out on ghost membranes. MetSO assay provided evidence of a significant increase in the oxidation of methionine from young to mature and senescent erythrocytes, whereas no significant difference appeared between mature and senescent cells (Fig. 2). The amount of MetSO in middle-aged erythrocytes is 50% higher than in membrane proteins of young cells, but it does not increase further. A possible ex-

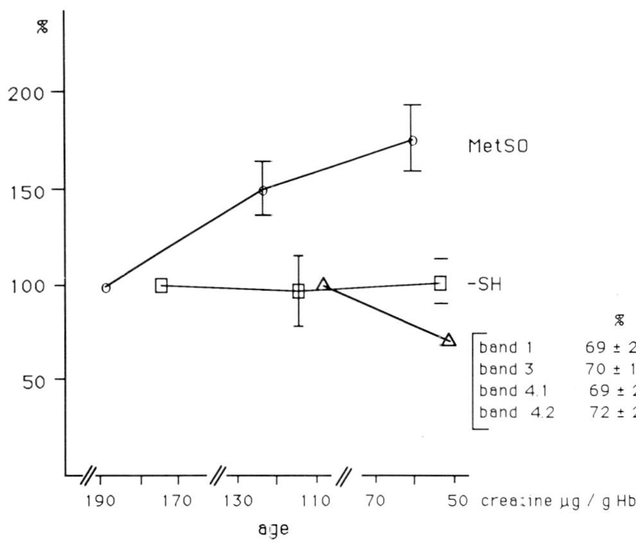

Fig. 2. Oxidative state of membrane proteins during red cell life span was quantitated by measuring methionine sulfoxide (*MetSO*) [54] and thiol group content [58] of ghost membranes obtained from erythrocytes of different ages, density separated by centrifugation in autologous plasma [53 – 55] and characterized for their creatine content as a cell-age parameter [56, 57]. Oxidative state of individual membrane proteins was determined in intact cells by measuring the amount of the thiol-reagent *N*-(7-dimethylamino-4-methylcoumarinyl)-maleimide (*DACM*) bound in situ by each membrane protein [58]

planation of this behavior could be the degradation of heavily oxidized proteins and the loss of membrane fragments containing the damaged material. In fact, the cell size decreases during red cell life [33], and loss of membrane fragments takes place [59]. Oxidation of methionine residues, already evident in middle-aged red cells, is expected to lead to conformational changes of proteins and partial loss of their function and could finally trigger their degradation. The possibility that methionine oxidation marks proteins for degradation was suggested to us by results obtained submitting red cells to nutrient deprivation experiments. Nutrient deprivation was carried out by incubating middle-aged red cells for 24 h in the absence of nutrients under two different sets of conditions [60]: in the presence of 0.5 m*M* ethylenediaminetetraacetic acid (EDTA): membrane fragmentation occurs with the release into the medium of vesicles containing only integral proteins, and no proteolysis of membrane proteins can be discerned either at the level of vesicles or on the membrane of incubated cells [60] (Fig. 3) and in the presence of 1 m*M* Ca^{2+} (released membrane fragments contain protein aggregates and proteolyzed membrane proteins; no proteolysis of membrane components in the membrane of incubated cells can be detected) [60] (Fig. 3). After nutrient deprivation, for each condition two types of samples were collected, one containing ghost membranes of incubated cells and the other containing, besides ghost membranes, cell debris and part of the released vesicles, and oxidation of membrane proteins was analyzed. While a generalized oxidation of thiols was observed under both conditions (not shown), comparing the MetSO content of ghosts prepared from incubated cells revealed that methionine appeared oxidized only in samples in which proteolysis was not active (Fig. 3). Where

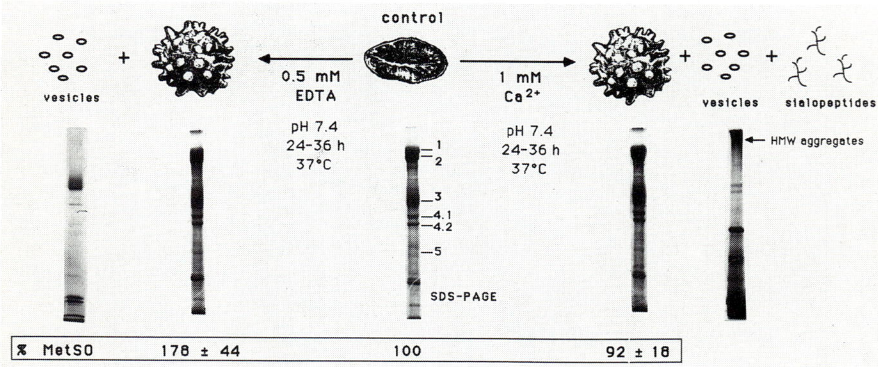

Fig. 3. Oxidation of methionine residues in membrane proteins in nutrient-deprived red cells. Middle-aged red cells were incubated in the absence of nutrients, in the presence of 0.5 m*M* ethylenediaminetetraacetic acid (*EDTA*) or 1 m*M* Ca^{2+}. Depicted are membrane processes occurring when red cells were nutrient deprived in the presence of 0.5 m*M* EDTA and 1 m*M* Ca^{2+}, as shown by SDS-PAGE carried out on ghost membranes prepared from incubated red cells and on membrane fragments released in the incubation medium and by chemical analysis of soluble material released in the incubation medium [60]. *HMW*, high molecular weight

proteolysis was active, oxidized methionine was confined to cell debris and membrane fragments released into the medium [61]. This result supports the hypothesis that MetSO marks proteins for proteolysis, and therefore assay of the MetSO content of membrane proteins could also monitor, besides oxidation, the loss of membrane material throughout the red cell life in the circulation.

The thiol content of proteins determined in ghost membranes appeared similar in all cell subpopulations studied (see Fig. 2). In order to obtain information about the thiol state of individual membrane proteins, their thiol content was assayed in intact cells by spectrofluorometry, using a coumarinyl derivative of maleimide (DACM) that after binding to free thiols becomes fluorescent [62]. This reagent crosses the membrane and binds to accessible thiols of integral and peripheral proteins. After incubation of young, middle-aged, and senescent cells with 3 m*M* DACM, ghost membranes were prepared and submitted to SDS-PAGE. Densitometry of fluorescent and Coomassie blue-treated gels allowed us to quantitate the amount of DACM bound to each membrane protein, which is indicative of the number of accessible thiol groups. The amount of DACM bound was quantitated by measuring the ratio between fluorescence intensity emitted by each protein and absorbance due to Coomassie blue staining [58]. After SDS-PAGE separation of ghost proteins under nonreducing conditions, 85 % of thiol groups were found labelled in the three red cell subpopulations studied [58]. Results obtained by calculating for each protein the ratio between fluorescence intensity and absorbance due to Coomassie blue staining showed that membrane proteins bind the same amount of DACM in young and mature erythrocytes [58]. On the contrary, in senescent red cells a significant decrease in the amount of DACM bound by band 3, α-spectrin, 4.1, and 4.2 proteins was evident [58] (see Fig. 2). Results obtained are indicative of conformational changes of band 3 and of the main proteins of the membrane skeleton in senescent cells, or of oxidation of these proteins, or of both these processes.

Functional Analysis of Band 3 Protein During Red Cell Life Span

Results obtained in studies concerning thiol distribution on individual membrane proteins in red cell subpopulations of different ages support the conclusion that conformational changes or oxidation of band 3 occurs in senescent cells. To investigate whether or not these processes affect the functional capacity of the protein in anion translocation, functional analysis of band 3 protein was carried out on density-separated, young, middle-aged, and senescent cells by measuring the binding to band 3 of three covalent fluorochrome inhibitors of anion exchange [63], 4,4'-diisothiocyanostilbene-2,2'-disulfonic acid disodium salt (DIDS) [64], eosine-5-maleimide (EM) [65, 66], and eosine-isothiocyanate (EITC) [67], directed to different subsites of the anion channel. Intact cells of different ages were treated with the fluorochrome, and then ghost membranes were prepared from the labelled cells and submitted to SDS-

PAGE. The amount of fluorochrome bound to band 3 was determined after densitometry of fluorescent gels and of the same gels after Coomassie blue treatment by measuring the ratio between fluorescence intensity emitted by the fluorochrome and absorbance due to Coomassie blue staining [63]. The quantitative results showed a decrease in the binding of DIDS and EITC in mature and senescent cells, and no differences in EM binding (Fig. 4). Although the effect of protein aggregation induced by the anion transport inhibition and differences in the labelling time course deserve further investigation, the results obtained show that a functional impairment of DIDS and EITC binding sites is already evident in middle-aged red cells. Methionine oxidation and functional impairment of these anion-binding subsites seem to occur in parallel. Oxidation of the band 3 protein molecule could be the earliest event leading to its functional impairment and finally to senescence antigen production.

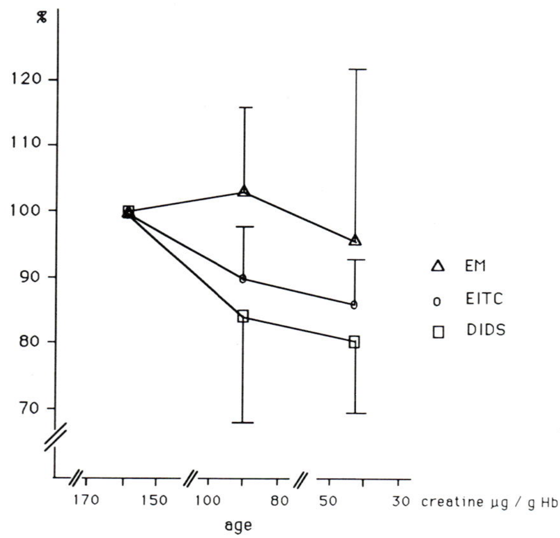

Fig. 4. Functional activity of band 3 protein as anion transporter during red cell life span, measured by quantitating the binding to band 3 of three covalent inhibitors of anion exchange [63]: 4-4'-diisothiocyanostilbene-2,2'-disulfonic acid disodium salt (*DIDS*) and eosine-5-maleimide (*EM*), that bind to two lysine residues [64–66], and eosine-isothiocyanate (*EITC*), that probably binds to a histidine [67] of the anion channel. Young, middle-aged, and senescent cells, characterized for their creatine content as a cell-age parameter, were treated with the fluorochrome [63]; ghost membranes were prepared from labelled cells and submitted to SDS-PAGE. The amount of fluorochrome bound to band 3 was determined after densitometry of fluorescent gels and of the same gels after Coomassie blue treatment by measuring the ratio between fluorescence intensity emitted by the fluorochrome and absorbance due to Coomassie blue staining [63]

Membrane Processes in Diamide-Treated Red Cells

Oxidation of red cells with diamide has been used as a model of the senescent cell, in particular to mimic cell aging at the membrane level [49]. Since diamide treatment induces a decrease of the reduced glutathione concentration [52], we hypothesized that after treatment of red cells with this reagent other phenomena besides thiol oxidation and membrane protein clustering could occur. Therefore, after treatment of a middle-aged red cell subpopulation with increasing concentrations of diamide, in addition to membrane protein oligomerization and thiol oxidation, oxidation of methionine to its sulfoxide form was analyzed in ghost membrane proteins. The presence of reducible protein aggregates was observed after SDS-PAGE separation in samples treated with 0.8 and 2.0 m*M* diamide, as already known from the literature. When red cells were treated with a fluorochrome that specifically binds to band 3 (EM), the membrane protein aggregates appeared fluorescent, thus indicating that band 3 participates in the formation of these protein clusters [63]. Quantitation of the thiol groups in ghost membranes obtained from diamide-treated red cells showed a significant decrease of free thiols only in samples treated with 2.0 m*M* diamide (Fig. 5). Furthermore, it is worth observing that a slight but significant increase in the amount of thiols can be seen in samples treated with 0.2 m*M* diamide. Oxidation of methionine residues is evident in membrane proteins of 0.2 and 2.0 m*M* diamide-treated cells (Fig. 5). Oxidation to MetSO appears to be produced under milder oxidative conditions (0.2 m*M* diamide) than those determining decrease of the thiol group content in membrane proteins (2.0 m*M* diamide). An increase of diamide concentration leads to a relative decrease of MetSO (Fig. 5), and since it is reasonable to suspect the pos-

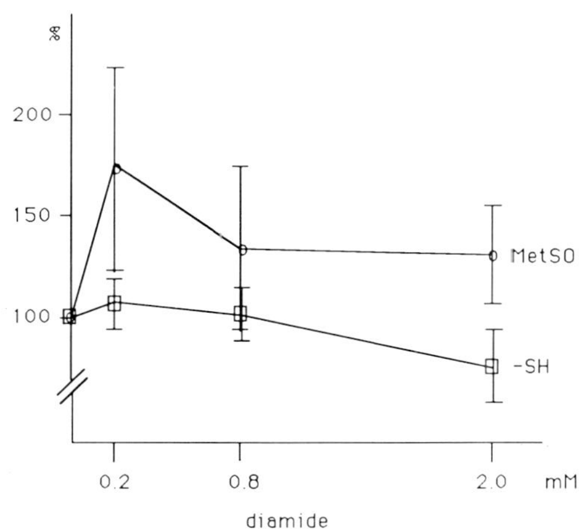

Fig. 5. Oxidative state of membrane proteins in diamide-treated red cells. Middle-aged red cells were treated with increasing amounts of diamide (0.2, 0.8, and 2.0 m*M*) [63]. The oxidative state of membrane proteins was determined by analysing the methionine sulfoxide (MetSO) and thiol content of ghost membranes prepared from diamide-treated erythrocytes

sibility of protein degradation and membrane fragmentation in diamide-treated red cells, this hypothesis is now under investigation.

Discussion

In searching for parameters to quantitate the oxidative state of membrane proteins in human erythrocytes during senescence, we observed that assay of the MetSO content of ghost membranes is a good indicator of the oxidative processes, since the MetSO content of membrane proteins is 50% higher in middle-aged and senescent red cells with respect to younger cells. Oxidation of methionine takes place early during the red cell life span in the circulation, since it is already evident in middle-aged cells and reaches nearly the same extent in middle-aged and senescent erythrocytes. The evaluation of the thiol group content of membrane proteins is not informative about oxidative processes when performed on ghost membranes obtained from different red cell subpopulations. However, when the thiol state of membrane proteins was evaluated in intact cells and individual membrane proteins considered, after SDS-PAGE separation a significant decrease of the amount of the thiol reagent DACM bound to band 3, α-spectrin, 4.1, and 4.2 proteins was observed in senescent cells. This can be indicative of oxidation or of conformational changes of these proteins in senescent erythrocytes; from these processes a different topology or topography of membrane proteins can result (as suggested by properties of red cells in some hemolytic anemias in which oxidative processes at the membrane level occur) [68 – 70].

Many membrane proteins are rich in cysteine and methionine residues. Oxidation of them could participate in a different way in the expression of membrane properties typical of the senescent red cell, the erythrocyte's antioxidant equipment is differently effective against the two lesions. A MetSO reductase, which reduces protein-bound MetSO to methionine, is present in numerous organisms and tissues [71, 72]. The presence of this enzyme has not been described in erythrocytes; therefore, oxidation of methionine cannot be repaired by red cells, other than by degradation of heavily oxidized proteins [73 – 75]. The accumulation of MetSO in middle-aged cells could trigger degradation of proteins and lead to local instability of the membrane, with loss of membrane fragments. Interestingly, comparing oxidative processes occurring in physiologically aged cells and in diamide-treated cells, oxidation of methionine to its sulfoxide form in the latter appears to be produced under milder oxidative conditions than those determining a decrease of thiol content in membrane proteins and during red cell life in the circulation can be discerned earlier than thiol modification. Red cells are equipped to maintain the reduced state of thiol groups through the glutathione system. In senescent cells, an impairment of the redox systems [16] could lead to oxidation of some thiols in membrane proteins. In fact, in senescent cells band 3, α-spectrin, 4.1, and 4.2 proteins seem to have a decreased thiol content or to have undergone conformational changes resulting in a decreased accessibility of their thiol groups to a thiol reagent. Band 3 protein is involved in the expression on the senescent cell surface of the antigen promoting IgG binding and clearance by macrophages [2, 3]. Functional analysis of band 3 showed that an impairment of DIDS and EITC binding sites is already expressed in middle-aged red cells. Methionine oxidation and a functional impairment of anion binding subsites seem to occur in parallel. Although at present we have no data about a possible relationship between oxidation of methionine and functional impairment of band 3, oxidation of this molecule could be the earliest event leading to the senescence antigen production.

Results obtained in these studies concerning the mechanisms of development of senescence markers during the red cell life span suggest several points. A comparison between the membrane properties of middle-aged and senescent cells is useful to study the presence of senescence markers on the cell surface that permit recognition of aged red cells for removal, and the thiol modification observed in senescent cells in our studies seems to be a senescence marker. In order to examine the mechanisms leading to the expression of senescence markers, studies about the membrane processes occurring throughout cell life in the circulation could give more information. Investigations carried out so far show that membrane fragmentation is a process occurring throughout cell life, and our studies have shown that oxidation of methionine and functional impairment of band 3 are already expressed in middle-aged cells. Oxidation and proteolysis of oxidized proteins are among the processes leading to this membrane fragmentation. Loss of membrane fragments is an aging process, and immunological mechanisms proposed for the clearance of senescent cells could also operate in the removal of membrane fragments lost by circulating red cells. A senescent cell binds IgG irreversibly and seems to be a cell no longer able to undergo loss of its plasma membrane. Why is this limit achieved? What happens on the cell surface when the cell is no longer able to repair its membrane by losing damaged regions? The answer to these questions would allow us to understand better how senescence markers develop.

References

1. Eadie GS, Brown IW (1953) Red blood cell survival studies. Blood 8: 1110–1136
2. Kay MMB (1975) Mechanism of removal of senescent cells by human macrophages 'in situ'. Proc Natl Acad Sci USA 72: 3521–3525
3. Kay MMB (1978) Role of physiologic autoantibody in the removal of senescent human red cells. J Supramol Struct 9: 555–567
4. Rice-Evans C, Omorphos SC, Baysal E (1986) Sickle cell pathology: is the membrane important? In: Rice-Evans C (ed) Free radicals, cell damage and disease. Richelieu, London, pp 149–166
5. Op den Kamp JAF (1979) Lipid asymmetry in membranes. Ann Rev Biochem 48: 47–71
6. Schlegel RA, McEvoy L, Weiser M, Williamson P (1987) Phospholipid organization as a determinant of red cell recognition by the reticuloendothelial system. In: Ropars C, Chassaigne M, Nicolau C (eds) Red blood cells as carriers for drugs: potential therapeutic applications, Pergamon, Oxford, pp 173–181

7. Viitala J, Järnefelt J (1985) The red cell surface revisited. Trends Biochem Sci 10: 392 – 395
8. Anstee DJ, Tanner MJA (1986) Structure and function of the red cell membrane sialoglycoproteins. Brit J Haematol 64: 211 – 215
9. Jay D, Cantley L (1986) Structural aspects of the red cell anion exchange protein. Ann Rev Biochem 55: 511 – 538
10. Marchesi VT (1985) Stabilizing infrastructure of cell membranes. Ann Rev Cell Biol 1: 531 – 561
11. Lazarides E (1987) From genes to structural morphogenesis: the genesis and the epigenesis of a red blood cell. Cell 51: 345 – 356
12. Gardner K, Van Bennett G (1989) Recently identified erythrocyte membrane-skeletal proteins and interactions: implications for structure and function. In: Agre P, Parker JC (eds) Red blood cell membranes. Dekker, New York, pp 1 – 30
13. Anderson RA (1989) Regulation of protein 4.1-membrane associations by a phosphoinositide. In Agre P, Parker JC (eds) Red blood cell membranes. Dekker, New York, pp 187 – 236
14. Rybicki AC, Heath R, Lubin B, Schwartz RS (1988) Human erythrocyte protein 4.1 is a phosphatidylserine binding protein. J Clin Invest 81: 255 – 260
15. Berk DA, Hochmuth RM, Waugh RE (1989) Viscoelastic properties and rheology. In: Agre P, Parker JC (eds) Red blood cell membranes. Dekker, New York, pp 423 – 454
16. Clark MR (1988) Senescence of red blood cells: progress and problems. Physiol Rev 68: 503 – 554
17. Lutz HU (1987) Red cell clearance (a review). Biomed Biochim Acta 46: S65 – S71
18. Kay MMB, Goodman SR, Sorensen K, Whitfield CF, Wong P, Zaki L, Rudloff V (1983) Senescent cell antigen is immunologically related to band 3. Proc Natl Acad Sci USA 80: 1631 – 1635
19. Kay MMB (1984) Localization of senescent cell antigen on band 3. Proc Natl Acad Sci USA 81: 5753 – 5757
20. Lutz HU, Flepp R, Stringaro-Wipf G (1984) Naturally occurring autoantibodies to exoplasmic and cryptic regions of band 3 protein, the major integral membrane protein of human red blood cells. J Immunol 133: 2610 – 2618
21. Galili U, Rachmilewitz EA, Peleg A, Flechner I (1984) A unique natural human IgG antibody with anti-alpha-galactosyl specificity. J Exp Med 160: 1519 – 1531
22. Galili U, Flechner I, Knyszynski A, Danon D, Rachmilewitz EA (1986) The natural anti-alpha-galactosyl IgG on human normal senescent red blood cells. Br J Haematol 62: 317 – 324
23. Galili U, Macher BA, Buehler J, Shohet SB (1985) Human natural anti-alpha-galactosyl IgG. II. The specific recognition of alpha(1-3)-linked galactose residues. J Exp Med 162: 573 – 582
24. Galili U, Basbaum CB, Shohet SB, Buehler J, Macher BA (1987) Identification of erythrocyte gal alpha 1-3gal glycosphingolipids with a mouse monoclonal antibody, Gal 13. J Biol Chem 262: 4683 – 4688
25. Clark MR, Sorette MP (1990) Characterization of antibody that binds in vivo to normal human red blood cells. International Symposium on Red Blood Cell Aging, Urbino, 24 – 26 September 1990, Abstract L28
26. Kay MMB, Goodman JR (1984) IgG antibodies do not bind to band 3 in intact erythrocytes; enzymatic treatment of cells is required for IgG binding. Biomed Biochim Acta 43: 841 – 846
27. Schweizer E, Angst W, Lutz HU (1982) Glycoprotein topology on intact human red blood cells reevaluated by cross-linking following amino group supplementation. Biochemistry 21: 6807 – 6818
28. Low PS, Waugh SM, Zinke K, Drenckhahn D (1985) The role of hemoglobin denaturation and band 3 clustering in red blood cell aging. Science 227: 531 – 533
29. Low PS (1989) Interaction of native and denatured hemoglobins with band 3: consequences for erythrocyte structure and function. In Agre P, Parker JC (eds) Red blood cell membranes. Dekker, New York, pp 237 – 260

30. Jancik J, Schauer R, Streicher MJ (1975) Influence of membrane-bound N-acetylneuraminic acid on the survival of erythrocytes in man. Hoppe Seyler's Z Physiol Chem 356: 1329 – 1331
31. Gattegno L, Perret G, Fabia F, Cornillot P (1981) Decrease of carbohydrate in membrane glycoproteins during human erythrocytes aging in vivo. Mech Ageing Dev 16: 205 – 219
32. Kempka G, Kolb-Bachofen V (1981) Galactose-specific receptors on liver cells. Hepatocyte and liver macrophage receptors differ in their membrane anchorage. Biochim Biophys Acta 847: 108 – 114
33. Westerman MP, Pierce LE, Jensen WN (1963) Erythrocyte lipids: a comparison of normal young and normal old populations. J Lab Clin Med 62: 394 – 400
34. Winterbourn CC, Batt RD (1970) Lipid composition of human red cells of different ages. Biochim Biophys Acta 202: 1 – 8
35. Cohen NS, Ekholm JE, Luthra MG, Hanahan DJ (1976) Biochemical characterization of density separated human erythrocytes. Biochim Biophys Acta 419: 229 – 242
36. Clark MR, Mohandas N, Shohet SB (1983) Osmotic gradient ektacytometry: comprehensive characterization of red cell volume and surface maintenance. Blood 61: 899 – 910
37. Nash GB, Meiselman HJ (1983) Red cell and ghost viscoelasticity. Effects of hemoglobin concentration and in vivo aging. Biophys J 43: 63 – 73
38. Weed RI (1970) The importance of erythrocyte deformability. Am J Med 49: 147 – 150
39. Williams AR, Morris DR (1980) The internal viscosity of the human erythrocyte may determine its life span in vivo. Scand J Haematol 24: 57 – 62
40. Snyder LM, Leb L, Piotrowski J, Sauberman N, Liu SC, Fortier NL (1983) Irreversible spectrin-haemoglobin crosslinking 'in vivo': a marker for red cell senescence. Brit J Haematol 53: 379 – 384
41. Campwala HQ, Desforges JF (1982) Membrane-bound hemichrome in density-separated cohorts of normal (AA) and sickled (SS) cells. J Lab Clin Med 99: 25 – 28
42. Jain SK (1988) Evidence for membrane lipid peroxidation during the 'in vivo' aging of human erythrocytes. Biochim Biophys Acta 937: 205 – 210
43. Morrison M, Au KS, Hsu L (1987) Are the red cell proteases a clock mechanism wich turns on a signal of senescence? Biomed Biochim Acta 46: S79 – S83
44. Sutera SP, Gardner RA, Boylan CW, Carroll GL, Chang KC, Marvel JS, Kilo C, Gonen B, Williamson JR (1985) Age-related changes in deformability of human erythrocytes. Blood 65: 275 – 282
45. Barber JR, Clarke S (1983) Membrane protein carboxyl methylation increases with human erythrocyte age – evidence for an increase in the number of methylatable sites. J Biol Chem 258: 1189 – 1196
46. Kay MMB, Bosman GJCGM, Lawrence C (1988) Functional topography of band 3: specific structural alteration linked to functional aberrations in human erythrocytes. Proc Natl Acad Sci USA 85: 492 – 496
47. Johnson GJ, Allen DW, Flynn TP, Finkel B, White JG (1980) Decreased survival 'in vivo' of diamide-incubated dog erythrocytes. J Clin Invest 66: 955 – 961
48. Kay MMB, Bosman GJCGM, Shapiro SS, Bendich A, Bassel PS (1986) Oxidation as a possible mechanism of cellular aging: vitamin E deficiency causes premature aging and IgG binding to erythrocytes. Proc Natl Acad Sci USA 83: 2463 – 2467
49. Arese P, Bussolino F, Flepp R, Stammler P, Fasler S, Lutz HU (1987) Diamide enhances phagocytosis of human red cell in a complement- and anti band 3 antibody-dependent process. Biomed Biochim Acta 46: S84 – S87
50. Beppu M, Mizukami A, Nagoya M, Kikugawa K (1990) Binding of anti-band 3 autoantibody to oxidatively damaged erythrocytes. J Biol Chem 265: 3226 – 3233
51. Oliver CN, Ahn B, Moerman EJ, Goldstein S, Stadtman ER (1987) Age-related changes in oxidized proteins. J Biol Chem 262: 5488 – 5491
52. Kosower NS, Kosower EM, Wertheim B (1969) Diamide, a new reagent for the intracellular oxidation of glutathione to the disulfide. Biochem Biophys Res Commun 37: 593 – 596

53. Beutler E, West C, Blume KG (1976) The removal of leukocytes and platelets from whole blood. J Lab Clin Med 88: 328 – 333
54. Brovelli A, Seppi C, Pallavicini G, Balduini C (1983) Membrane processes during 'in vivo' aging of human erythrocytes. Biomed Biochim Acta 42: S122 – S126
55. Murphy JR (1973) Influence of temperature and method of centrifugation on the separation of erythrocytes. J Lab Clin Med 82: 334 – 341
56. Griffiths WJ (1964) The determination of creatine in body fluids and muscle, and of phosphocreatine in muscle, using the autoanalyzer. Clin Chim Acta 9: 210 – 213
57. Fehr J, Knob M (1979) Comparison of red cell creatine level and reticulocyte count in appraising the severity of hemolytic processes. Blood 53: 966 – 976
58. Seppi C, Castellana MA, Minetti G, Piccinini G, Balduini C, Brovelli A (1991) Evidence for membrane protein oxidation during 'in vivo' aging of human erythrocytes. Mech Ageing Dev 57: 247 – 258
59. Morrison M, Michaels AW, Phillips DR, Choi S (1974) Life span of erythrocyte membrane protein. Nature 248: 763 – 789
60. Brovelli A, Seppi C, Bardoni A, Balduini C, Lutz HU (1987) Re-evaluation of the structural integrity of red-cell glycoproteins during aging 'in vivo' and nutrient deprivation. Biochem J 242: 115 – 121
61. Brovelli A, Seppi C, Castellana MA, De Renzis MR, Blasina A, Balduini C (1990) Oxidative lesion to membrane proteins in senescent erythrocytes. Biomed Biochim Acta 49: S218 – S223
62. Yamamoto K, Sekine T, Kanaoka Y (1977) Fluorescent thiol reagents – Fluorescent tracer method for protein SH groups using *N*-(7-dimethylamino-4-methyl coumarinyl) maleimide. An application to the proteins separated by SDS-polyacrylamide gel electrophoresis. Anal Biochem 79: 83 – 94
63. Brovelli A, Castellana MA, Minetti G, Piccinini G, Seppi C, De Renzis MR, Balduini C (1990) Conformational changes and oxidation of membrane proteins in senescent human erythrocytes. International symposium on red blood cell aging. Urbino. 24 – 26 September 1990 (in press)
64. Ho MK, Guidotti G (1975) A membrane protein from human erythrocytes involved in anion exchange. J Biol Chem 250: 675 – 683
65. Nigg E, Kessler M, Cherry RJ (1979) Labelling of human erythrocyte membranes with eosin probes used for protein diffusion measurements. Inhibition of anion transport and photo-oxidative inactivation of acetylcholinesterase. Biochim Biophys Acta 550: 328 – 340
66. Cobb CE, Beth AH (1990) Identification of the eosinyl-5-maleimide reaction site on the human erythrocyte anion exchange protein: overlap with the reaction sites of other chemical probes. Biochemistry 29: 8283 – 8290
67. Chiba T, Sato Y, Suzuki Y (1987) Characterization of eosin 5-isothiocyanate binding site in band 3 protein of the human erythrocyte. Biochim Biophys Acta 897: 14 – 24
68. Rank BH, Carlsson J, Hebbel RP (1985) Abnormal redox status of membrane-protein thiols in sickle erythrocytes. J Clin Invest 75: 1531 – 1537
69. Becker PS, Morrow JS, Lux SE (1987) Abnormal oxidant sensitivity and beta chain structure of spectrin in hereditary spherocytosis associated with defective spectrin binding. J Clin Invest 80: 557 – 565
70. Arese P, De Flora A (1990) Pathophysiology of hemolysis in glucose-6-phosphate dehydrogenase deficiency. Semin Hematol 27: 1 – 40
71. Brot N, Weissbach L, Werth J, Weissbach H (1981) Enzymatic reduction of protein bound methionine sulfoxide. Proc Natl Acad Sci USA 78: 2155 – 2158
72. Spector A, Scotto R, Weissbach H, Brot N (1982) Lens methionine sulfoxide reductase. Biochem Biophys Res Commun 108: 429 – 434
73. Levine RL, Oliver CN, Fulks RM, Stadtman ER (1981) Turnover of bacterial glutamine synthetase: oxidative inactivation precedes proteolysis. Proc Natl Acad Sci USA 78: 2120 – 2124
74. Rivett AJ (1985) Preferential degradation of the oxidatively modified form of glutamine synthetase by intracellular mammalian proteases. J Biol Chem 260: 300 – 305
75. Davies KJA, Goldberg AL (1987) Proteins damaged by oxygen radicals are rapidly degraded in extracts of red blood cells. J Biol Chem 262: 8227 – 8234

Aging and Electron Leakage Outside Mitochondrial Energy-Transducing Sequence

G. Benzi and D. Curti

Introduction

The mitochondrial electron transfer from donors (reduced nicotinamide-adenine dinucleotide, NADH, and succinate) to molecular oxygen results in the vectorial translocation of protons across the inner membrane. This proton electrochemical potential ($\Delta\mu H^+$) drives ATP synthesis, ion translocation, and protein importation. At three steps along the electron transfer chain, represented by protein-lipid enzyme complexes I, III, and IV, oxidative energy is conserved via this generation of protonic energy (Fig. 1).

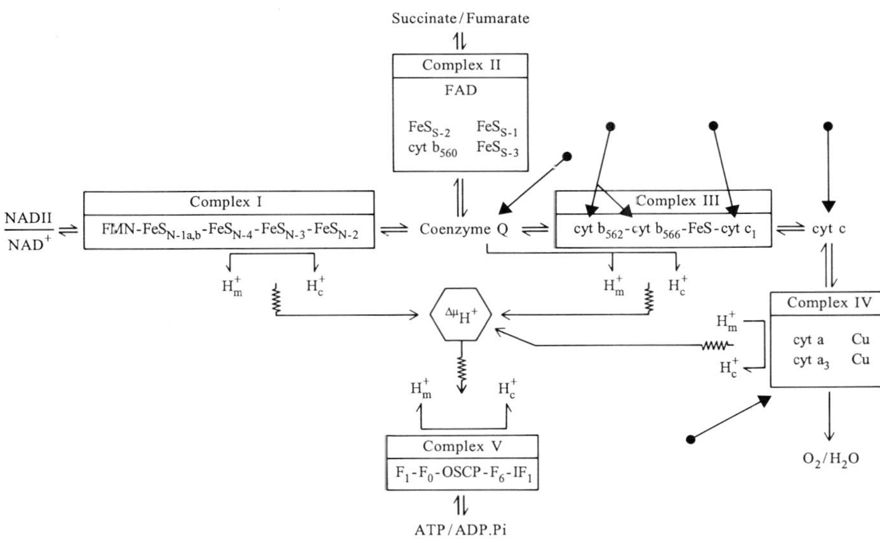

Fig. 1. Scheme of the mitochondrial electron transfer oxidative phosphorylation system depicting the energy communication by way of the electrochemical potential of protons ($\Delta\mu_{H^+}$) among the complexes (I, III, IV, and V). Iron-sulfur cluster (FeS) subscripts indicate the individual compounds, while subscripts on cytochromes *b* indicate the position of their α peaks at ambient temperatures. H^+_m and H^+_c indicate protons on the matrix and cytosolic sides, respectively of the mitochondrial inner membrane. *Arrows* indicate the respiratory components tested in the present research. *FMN*, flavin mononucleotide; *FAD*, flavin adenine dinucleotide; *OSPC*, oligomycin sensitivity-conferring protein; F_b, coupling factor b; IF_1, ATPase inhibitor protein; F_1, sector headpiece; F_0, sector basepiece

Reihe der Villa Vigoni, Band 1
Biology of Aging
Zwilling and Balduini (Eds.)

The electron carriers of the mitochondrial respiratory chain (the complexes plus ubiquinone and cytochrome *c*) are quinoid structures (flavin mononucleotide , FMN, flavin adenine dinucleotide, FAD, coenzyme Q) or transition metal complexes (iron-sulfur clusters, hemes, protein-bound copper). In the quasi-equipotential regions (represented by the complexes), there would be minimum energy loss and maximum energy conservation. The system up to cytochrome *c* is reversible, while the terminal step from cytochrome aa_3 to oxygen is not, displacing the equilibrium of the system in the direction of ATP synthesis.

Although it has long been suggested that cerebral aging may be due to damage caused by free radicals produced in the electron transfer chain, the mechanism causing control of the mitochondrial electron flow from electron donors to cytochrome aa_3 to be lost during aging has not yet been clarified.

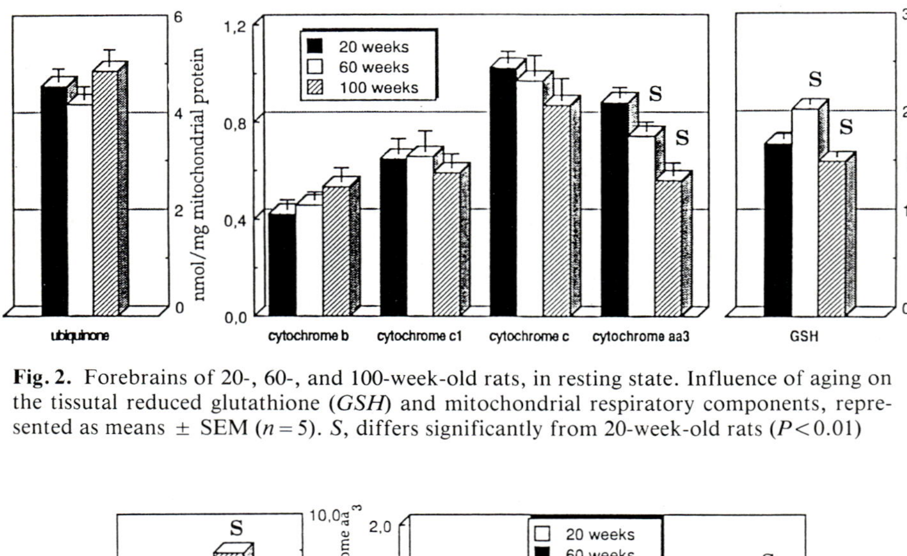

Fig. 2. Forebrains of 20-, 60-, and 100-week-old rats, in resting state. Influence of aging on the tissutal reduced glutathione (*GSH*) and mitochondrial respiratory components, represented as means ± SEM ($n=5$). *S*, differs significantly from 20-week-old rats ($P<0.01$)

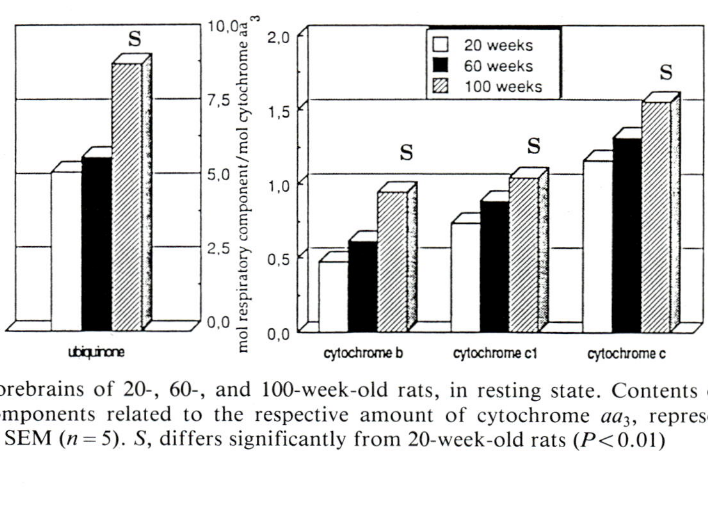

Fig. 3. Forebrains of 20-, 60-, and 100-week-old rats, in resting state. Contents of respiratory components related to the respective amount of cytochrome aa_3, represented as means ± SEM ($n=5$). *S*, differs significantly from 20-week-old rats ($P<0.01$)

Neurons are protected by an effective network of antioxidant mechanisms, and at cerebral level reduced glutathione (GSH) plays a pivotal role in protecting cells from oxidative stress.

According to our data, brain aging is characterized by the depletion of GSH concomitant with the decrease in the activity of cytochrome aa_3, while other key components of the electron transfer chain, namely ubiquinone and cytochrome *b* populations, remain practically unchanged when their content is expressed per mitochondrial protein (Fig. 2). This indicates that the calculated stoichiometry of the electron carriers in respect to cytochrome aa_3 shows an age-dependent increase in the proportionalities of the ubiquinone and cytochrome *b* populations (Fig. 3).

Role of Ubiquinone and Cytochrome *b* Populations in the Age-Related Electron Leakage

Both ubiquinone (or coenzyme Q) and cytochrome b_{566} (which is part of complex III; Fig. 1) could be oxygen free radical "producers" through an autooxidation process affecting their reduced forms. However, the empirical demonstration of their involvement in the process of oxygen radical formation is far from easy, since the metabolic conditions suitable for generating radicals also highly reduce both molecules [8, 10, 26, 27, 33]. Therefore, it is difficult to decide whether or not the reduced molecular species of ubiquinone, or cytochrome b_{566}, or both, are markedly involved in the formation of the oxygen free-radicals in the mitochondria.

The mitochondrial respiratory chain transfers electrons by the sequential interaction of carriers undergoing interconversion between reduced and oxidized states. Because of the kinetic spin restriction, oxygen prefers to accept electrons one at a time, involving the production of partially reduced oxygen intermediates, e.g., superoxide radical, hydrogen peroxide, hydroxyl radical, etc. (Fig. 4). The cytochrome oxidase of aa_3 type retains all the activated oxygen intermediates tightly bound to its active site until oxygen is completely reduced to water (oxygen with four electrons). In contrast, other autooxidizing components in the respiratory chain can transfer electrons directly into oxygen but do not retain the partially reduced oxygen intermediates bound to their active sites. Thus, activated oxygen species may be released.

The normal electron transfer from donors (NADH and succinate) to cytochrome aa_3 may be deviated under a state of high reduction of the components related to the zero potential pool [8, 25, 26, 33]. Ubiquinone and cytochrome *b* populations are postulated to release oxygen free radicals by autooxidation of their reduced forms. This hypothesis is based on many observations. Depletion of ubiquinone enables mitochondria to produce hydrogen peroxide, while ubiquinone reincorporation leads to a recovery of succinate-mediated respiration with a linear dose action relationship. Further evidence comes from the requirement for ubiquinone to lead to superoxide and hydrogen peroxide formation by the system NADH: ubiquinone re-

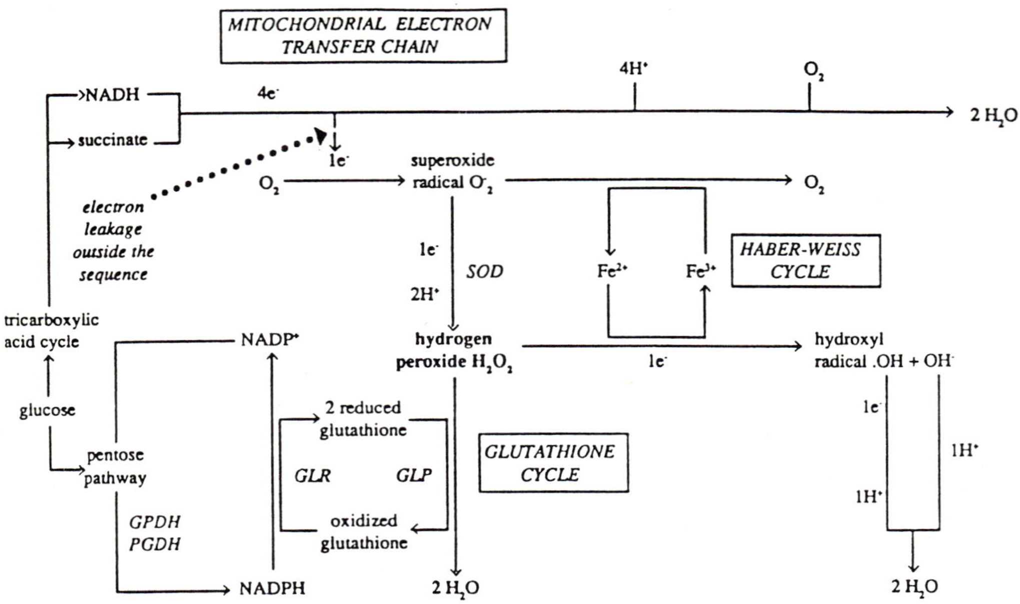

Fig. 4. Pivotal role of the glutathione cycle in protecting neurons from electron leakage and consequent oxygen free radical generation. Glutathione cycle antagonizes the Haber-Weiss cycle in which superoxide radical (O_2^-) furnishes hydrogen peroxide (H_2O_2) and regenerates Fe^{2+}, the two partners of the reaction which produces the strongly dangerous hydroxyl radical (.OH). *SOD*, superoxide dismutase; *GLP*, glutathione peroxidase; *GLR*, glutathione reductase; *GPDH*, glucose 6-phosphate dehydrogenase; *PGDH*, 6-phosphogluconate dehydrogenase

ductase [complex I]-ubiquinol-cytochrome *c* reductase [complex III] [9, 10, 26, 27, 33].

These observations should be regarded, however, with caution because: (a) the mitochondrial ubiquinone-extraction procedure may alter both the dehydrogenases [22] and the membrane phospholipids [43, 50]; (b) the amount of ubiquinone needed to restore electron flux fully is twice the normal content [36]; (c) excess ubiquinone added for the mitochondrial respiration reconstitution may be incorporated into the membrane aqueous phase where autooxidizing properties, redox power, etc. markedly differ from the biophysical conditions of the lipophilic structure in which ubiquinone physiologically operates [14, 41, 30].

Other observations confirm a direct relationship between the changes in the steady state of ubiquinone populations and the rates of oxygen radical production [10, 44]. The steady state concentration of ubisemiquinone remains, however, unchanged under experimental conditions [34] under which myxothiazol (inhibitor of the mitochondrial respiration at the site of cytochrome b_{566}) induces a nearly complete inhibition of superoxide formation in mitochondria respiring with succinate and poisoned with antimycin A, an antibiotic that shifts to negative values the redox potentials of the electron carriers ranging around zero (e.g., ubiquinone populations) and, therefore, causes a marked release of superoxide and hydrogen peroxide. These two apparently contradictory observations taken together suggest that the oxygen radical production may be driven not only by the redox cycle linked to the ubiquinone population but also by reduced forms of the cytochrome *b* population.

The possible molecular events associated with the mitochondrial generation of oxygen radicals may thus already be related to the normal processes of mitochondrial energy transduction, with both candidates involved in electron leakage (ubiquinone and cytochrome b_{566}) being strictly linked to electron flow through the respiratory chain complexes. In fact, cytochrome b_{566} is strictly correlated with the processes of energy transduction in Complex III, wavering continuously [12] between a very low potential state ($\xi'_0 = -30\,\text{mV}$) and a very high one ($\xi'_0 = +245\,\text{mV}$). This considerable variation in free energy is not, however, due to the intrinsic property of cytochrome b_{566} associated with the redox state but is prompted by the membrane potential, which is consolidated by the energy conservation of coupled mitochondria [7].

Cytochrome b_{566}'s low potential may play a primary role in the formation of mitochondrial superoxide radical, since an increase in its redox potential inhibits the univalent transfer of electrons to oxygen, and the presence of molecular species with a potential of $-30\,\text{mV}$ is paralleled by the release of superoxide radicals. Therefore, a leakage of electrons at the cytochrome b_{566} level would appear to be a realistic element, already accompanying the electron flow of coupled mitochondria and which can thus bring about a continual release of oxygen radicals. During aging, the small quantities of these radicals that manage to escape the local defense mechanisms (i.e., operated by the glutathione system) may lead to a multiplicity of changes in the chemical and physical state of the membranes, including the mitochondrial inner membrane.

In the above-quoted, age-related, chemicophysical alterations, the ubiquinone population can produce oxygen free radicals. In fact, under normal conditions ubiquinone is arranged in the mitochondrial membrane in such a way that a hydrophilic quinone residue is situated in the hydrophobic environment [11]. Here, the redox state of ubiquinone could be similar to its behavior in an aprotic solvent. The reaction of the ubisemiquinone radical with oxygen is an exergonic event which is active in aqueous solutions. On the contrary, the autooxidizing process cannot take place at a significant rate in a hydrophobic medium [34]. This situation would appear to be the same as that normally found in the mitochondria because of the absence of protons in the double phospholipid layer of the mitochondrial inner membrane. The formation of superoxide radicals in the mitochondrial membrane cannot arise through the autooxidizing ubisemiquinone process since the protons are prevented from penetrating the hydrophobic domain in which ubiquinone is in an operative state.

An age-related increase in the electron leakage at the level of cytochrome b_{566} produces the above-quoted release of oxygen free radicals, inducing an alteration to the structure of the mitochondrial inner membrane. This alteration affects the double layer of phospholipids, thus increasing the hydrophilic properties of the mitochondrial inner membrane, an event forming an endless loop which creates the thermodynamic conditions favoring the destabilization of ubisemiquinones by autooxidizing processes that easily occur in aqueous phases [21, 31, 33].

The Age-Related Decrease in Cytochrome aa_3 and Glutathione Concentrations as Promoter of Both Electron Leakage and Peroxidation

According to our data, aging is characterized by: (a) marked decrease in the brain GSH concentration, and (b) marked decrease in the mitochondrial content of cytochrome aa_3 without any change in the content of ubiquinone and cytochrome *b* populations. Thus, the decrease in activity of cytochrome oxidase of aa_3 type [1, 3, 20] may account for the fact that, when stoichiometric calculations are made, aging is characterized by a relative increase in the percentage of ubiquinone and cytochrome *b* populations. Although the amount of this imbalance between the electron carriers is not dramatic in any absolute sense [32], its existence in a relative sense (as revealed by stoichiometric calculations) does explain how electrons can „escape“ outside the electron transfer sequence from electron donors (NADH and succinate) to molecular oxygen (Fig. 5). In fact, only cytochrome aa_3 retains all the partially reduced oxygen intermediates bound to its active site, while ubiquinone and cytochrome *b* populations transfer the electron to oxygen but do not retain the partially reduced oxygen intermediates in their active sites.

The production of superoxide radical by electron leakage at the level of the cytochrome *b* population causes mitochondrial membrane alteration because of the dismutase reaction in which hydrogen peroxide is formed and which (because of the intervention of the Haber-Weiss cycle) leads to highly dangerous hydroxyl radicals (see Fig. 4). This condition might change the physicochemical membrane properties, favoring autooxidation of ubisemiquinones in mitochondria from old rats. Thus, increasing the leakage of electrons outside the sequence at the level of cytochrome *b* population increases the generation rates of partially reduced oxygen intermediates due to an age-related development of an additional electron donor for oxygen out of sequence, namely the ubiquinone population (Fig. 5).

The redox-cycling ubiquinone radicals are involved in the generation of hydroxyl radicals by mitochondria as demonstrated by the three- to sevenfold increase in hydroxyl radical formation in the presence of antimycin A, which could be completely abolished by thenoyltrifluoroacetone, a compound that inhibits mitochondrial respiration at the substrate site of complex II. A similar effect of antimycin A and thenoyltrifluoroacetone is present with respect to steady state concentrations of mitochondrial ubisemiquinone [35].

The essence of these observations is that ubisemiquinone can directly donate electrons to hydrogen peroxide, thus forming the hydroxyl radical

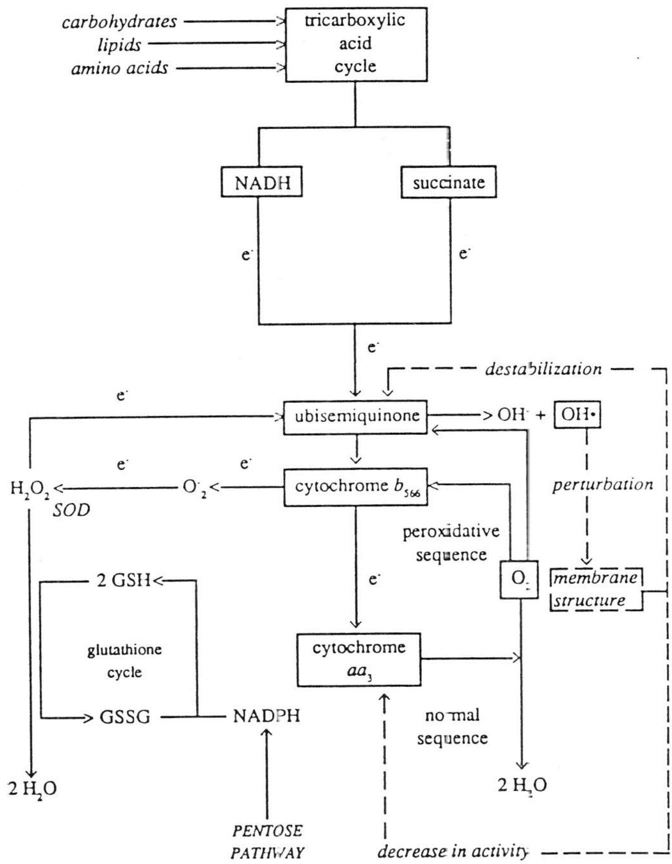

Fig. 5. Relationship between electron leakage at level of the cytochrome *b* population and the antiperoxidative glutathione system. *SOD*, superoxide dismutase; *GSH*, reduced glutathione; *GSSG*, oxidized glutathione

(Fig. 5). This fact is important in that hydrogen peroxide can permeate biological membranes (including the mitochondrial membrane) with extreme ease. Therefore, whatever the source of production of hydrogen peroxide, it can easily cross the mitochondrial inner membrane and come into contact with the ubisemiquinone radical, thus being transformed into the hydroxyl radical. Many of the hydrogen peroxide molecules react with GSH (see Fig. 4) and are thus detoxified and eliminated as water. However, because of the age-related decrease in GSH, increasing amounts of hydrogen peroxide not detoxified by glutathione can cause an increase in the production of hydroxyl radicals through the ubiquinone population (Fig. 5).

The leakage of electrons outside the normal energy-transducing sequence leads to the formation of superoxide radicals and highlights the inability of

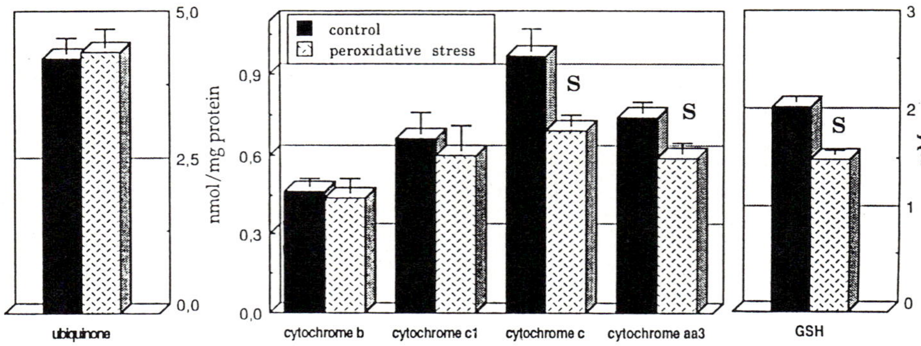

Fig. 6. Forebrains of 60-week-old rats analyzed 18 h after the intraperitoneal injection of the prooxidant cyclohexene-1-one (200 mg/kg). The tissutal reduced glutathione (GSH) and the mitochondrial respiratory components are represented as means ± SEM ($n = 5$). *S*, differs significantly from control rats not submitted to peroxidative stress ($P < 0.01$)

mitochondria to control electron flow efficiently through the components of the respiratory chain itself. In older animals, the release of superoxide radical and the dismutation product (hydrogen peroxide) could be a process of considerable proportions, and the leakage of electrons (as a mandatory side effect of the normal flux of electrons from both NADH and succinate to molecular oxygen) may substantially be due to changes in the availability of glutathione and in the balance between the components in the respiratory chain associated with the energy-transducing system.

The relationships between free radicals, GSH depletion, and imbalance between the components in the synaptic mitochondrial electron transfer chain are further supported by the observation that in the forebrains of 60-week-old rats submitted to peroxidative stress by cyclohexene-1-one there is an acute drop of GSH without any change in the contents of the ubiquinone and cytochrome *b* populations. In contrast, a marked decrease in the content of cytochrome aa_3 is present, just as occurs under the cerebral biochemical conditions induced by aging (Fig. 6).

Factors Involved in the Age-Related Modification of the Cytochrome aa_3 Content

The age-related decrease in the cytochrome oxidase activity of aa_3 type may be induced by the decrease either in the number of cytochrome aa_3 molecules or in the activity of each molecule of cytochrome aa_3. In synaptic mitochondria, the K_m for cytochrome *c* is practically the same in young and old rats, while the V_{max} of mitochondrial preparations from old animals is markedly lower than that in preparations from young ones. These findings indicate an age-dependent decrease in the number of cytochrome aa_3 molecules [15]. Furthermore, whereas substrates and enzymes of the glutathione cycle are related to the cytosolic compartment controlled by nuclear DNA, the location of the prosthetic groups of cytochrome oxidase of aa_3 type is confined to subunits I and II [48], which are encoded by mitochondrial DNA and synthesized within the mitochondria [45]. Aging decreases the turnover of mitochondrial DNA [23] and changes its structure and replication, with an increase in the frequency of circular dimers [39].

On the basis of many observations [16, 18, 24, 28, 29, 37, 42, 46, 47, 51], it may be reasonably posited that gene expression is modified during the life span, and this may be the result of alterations in control mechanisms rather than actual damage to the genome. All the quoted findings taken together indicate that the age-related imbalance between the components of the electron transfer chain and the age-associated decrease in GSH availability are modulated by changes in control of differently localized DNA. Thus, it is markedly difficult for exogenous interventions, if there are any, to modulate concomitantly the gene expression changes for both cytochrome aa_3 (encoded by mitochondrial DNA) and GSH (encoded by nuclear DNA).

In any case, the modification of one or more parameters by exogenous factors may be usefully investigated. In fact, concerning energy metabolism, aging is characterized by a decrease in ATP content and synthesis [19]. Furthermore, in synaptosomes isolated from old hypoxic brains, the free energy utilized for ATP synthesis is progressively lower than that released upon the transfer of electrons from NADH to cytochrome *c* [4 – 6]. Concerning the peroxidative processes, evidence is accumulating for the involvement of oxygen free radicals in neuronal damage during ischemia [17] and/or postischemic recirculation [40], as well as in idiopathic and toxin-induced parkinsonism in which the decrease in GSH in discrete brain regions plays a pivotal role [2, 13, 38, 49].

Conclusion

It is possible to conclude (see Fig. 5) that the mitochondrial membranes, where the ubiquinone population normally operates, can undergo variations in their chemical and physical state by electron leakage at the level of the autooxidizing cytochrome *b* population, inducing superoxide radical release. This, in turn, causes changes in the redox cycle of the ubiquinone population in such a way that it becomes autooxidizing. Ubiquinone's sensitivity may derive from its special function within the respiratory chain. Indeed, it is well-known that ubiquinone represents the only nonprotein carrier in the electron transfer system that functions by moving freely within the phospholipid double layer of the mitochondrial inner membrane, inasmuch as it is linked to specific proteins which ensure the directionality of the electron flow and the feasibility of proton translocation.

The function of this redox cycle is related to a cyclic operation of the ubiquinone population with freedom of movement in the mitochondrial inner membrane, unlike the other redox components in the respiratory chain.

Therefore, the ubiquinone population can react to chemical and physical inner membrane alterations with changes in its thermodynamic and kinetic properties towards oxygen. These changes in the membrane can account for the formation of oxygen free radicals, because of autooxidation of the ubiquinone population. Whatever the case, this depends also on adequate protonation of the membrane zone in which these molecules are located.

In some neuronal cells, aging affects these various parameters because of the gradual imbalance between the carriers of the electron transfer chain. Indeed, when calculating the stoichiometry of these carriers with respect to cytochrome aa_3, aging leads to an increase in the proportionalities of ubiquinone and cytochrome *b* populations. This results in the continuous release of partially reduced oxygen intermediates via both the ubiquinone and cytochrome *b* populations. Because of the age-related decrease in the cerebral GSH content increasing amounts of oxygen free radicals not detoxified by glutathione cause mitochondrial inner membrane damage which further stimulates their generation. In this way, the mitochondrial condition not only increases the leakage of oxygen radicals as a function of aging itself but may induce functional and structural changes until neuron death occurs.

Acknowledgements. We thank Gemma Garlaschi, Gianfranca Corbellini, and Domenica Minisci for their secretarial work. The technical assistance of Luigi Maggi and Giorgio Coscia is gratefully acknowledged.

References

1. Abu-Erreish GM, Sanadi DR (1978) Age-related changes of cytochrome concentration of myocardial mitochondria. Mech Ageing Dev 7: 425 – 432
2. Bannon MJ, Goedert M, Williams B (1984) The possible relation of glutathione, melanin and 1-methyl-4-phenyl-1,2,5,6-tetrahydropyridine (MPTP) to Parkinson's disease. Biochem Pharmacol 33: 2697 – 2698
3. Benzi G (1979) Enzymatic activities related to energy transduction in the mature rat brain. Interdiscipl Topics Geront 15: 104 – 120
4. Benzi G, Giuffrida AM (1985) Bioenergetics of hypoxic brain during aging. Mol Physiol 8: 535 – 547
5. Benzi G, Giuffrida AM (1987) Changes of synaptosomal energy metabolism induced by hypoxia during aging. Neurochem Res 12: 149 – 157
6. Benzi G, Agnoli A, Giuffrida AM (1984) Influence of aging on cerebral derangement by acute severe hypoxia during hypovolemic hypotension. Neurobiol Aging 5: 213 – 220
7. Berry EA, Trumpower BL (1985) Pathways of electrons and protons through the cytochrome bc_1 complex of the mitochondrial respiratory chain. In: Lenaz G (eds) Coenzyme Q. Wiley, New York, pp 365 – 389
8. Boveris A, Oshino N, Chance B (1972) The cellular production of hydrogen peroxide. Biochem J 128: 617 – 630
9. Boveris A, Cadenas E, Stoppani AOM (1976) Role of ubiquinone in the mitochondrial generation of hydrogen peroxide. Biochem J 156: 435 – 444
10. Cadenas E, Boveris A, Ragan CI, Stoppani AOM (1977) Production of O_2 radicals and H_2O_2 by NADH-ubiquinone reductase and ubiquinolcytochrome *c* reductase from beef heart mitochondria. Arch Biochem Biophys 180: 248 – 257
11. Chance B (1972) On probe and ubiquinone interactions in mitochondrial membranes. In: Azzone GF, Carafoli E, Lehninger AL, Quagliariello E and Siliprandi N (eds) Biochemistry and biophysics of mitochondrial membranes. Academic, New York, pp 85 – 99

12. Chance B, Wilson DF, Dutton PL, Erecinska M (1970) Energy coupling mechanisms in mitochondria: kinetic, spectroscopic, thermodynamic properties of an energy-transducing form of cytochrome *b*. Proc Natl Acad Sci USA 66: 1175 – 1182
13. Cohen G (1983) Catalase, glutathione peroxidase, superoxide dismutase, and cytochrome P-450. In: Lajtha A (ed) Handbook of neurochemistry, vol 4. Plenum, New York, pp 315 – 330
14. Crane FL (1977) Hydroquinone dehydrogenase. Annu Rev Biochem 46: 439 – 469
15. Curti D, Giangaré MC, Redolfi ME, Fugaccia I, Benzi G (1990) Age-related modifications of cytochrome *c* oxidase activity in discrete brain regions. Mech Ageing Dev 55: 171 – 180
16. Cutler RG (1984) Antioxidants, aging and longevity. In: Pryor WA (ed) Free radicals in biology, vol 6. Academic, New York, pp 371 – 428
17. Demopoulos H, Flamm E, Seligman M, Power R, Pietronigro D, Ransohoff J (1977) Molecular pathology of lipids in CNS membranes. In: Jöbsis FF (ed) Oxygen and physiological function. Professional Information Library, Dallas, pp 491 – 508
18. Florine DL, Ono T, Cutler RG, Getz MJ (1980) Regulation of endogenous murine leukemia virus-related nuclear and cytoplasmic RNA complexity in C57BL/6J mice with increasing age. Cancer Res 40: 519 – 523
19. Foris G, Leovey A (1980) Age related changes in cAMP and cGMP levels during phagocytosis in human polymorphonuclear leukocytes. Mech Ageing Dev 27: 233 – 237
20. Gold PH, Gee MW, Strehler BL (1968) Effect of age on oxidative phosphorylation in rat. J Gerontol 23: 509 – 512
21. Grinna LS (1977) Age-related changes in the lipids of the microsomal and the mitochondrial membranes of rat liver and kidney. Mech Ageing Dev 6: 197 – 205
22. Gutman M, Coles CJ, Singer TP, Casida JE (1971) On the functional organization of the respiratory chain at the dehydrogenase-coenzyme Q junction. Biochemistry 10: 2036 – 2043
23. Heumer RP, Lee KD, Reeves AE, Bickert C (1971) Mitochondrial studies in senescent mice. II. Specific activity, buoyant density and turnover of mitochondrial DNA. Exp Gerontol 6: 327 – 334
24. Le Stourgeon WM, Totten R, Forer A (1974) The nuclear acidic proteins in cell proliferation an differentiation. In: Cameron IL and Jeter JR (eds) Acidic proteins of the nucleus. Academic, New York, pp 159 – 190
25. Loschen G, Flohé L, Chance B (1971) Respiratory chain linked H_2O_2 production in pigeon heart mitochondria. FEBS Lett 18: 261 – 264
26. Loschen G, Azzi A, Flohé L (1973) Mitochondrial H_2O_2 formation: relationship with energy conservation. FEBS Lett 33: 84 – 88
27. Loschen G, Azzi A, Flohé L (1974) Mitochondrial hydrogen peroxide formation. In: Thurman RG, Yonetani T, Williamson JR and Chance B (eds) Alcohol and aldehyde metabolizing systems. Academic, New York, pp 215 – 229
28. Medvedev ZA, Medvedev MN, Robson L (1980) Age-related changes in the electrophoretic pattern of the high molecular weight non-histone proteins from mouse liver, hepatoma and spleen chromatin. Age 3: 74 – 77
29. Miquel J, Fleming JE (1986) Theoretical and experimental support for an "oxygen radical-mitochondrial injury" hypothesis of cell aging. In: Johnson JE Jr, Walford R, Harman D and Miquel J (eds) Free radicals, aging, and degenerative diseases. (Modern aging research, vol 8) Liss, New York, pp 51 – 74
30. Morison LE, Schelhorn JE, Cotton TM, Bering CL, Loach PA (1982) Electrochemical and spectral properties of ubiquinone and synthetic analogs: relevance to bacterial photosynthesis. In: Trumpower BL (ed) Function of quinones in energy conserving systems, Academic, New York, pp 35 – 58
31. Nohl H (1971) Influence of age on thermotropic kinetics of enzymes involved in mitochondrial energy-metabolism. Z Gerontol 12: 9 – 18
32. Nohl H (1986) Oxygen radical release in mitochondria: influence of age. In: Johnson JE Jr, Walford R, Harman D, Miquel J (eds) Free radicals, aging, and degenerative diseases (Modern aging research, vol 8) Liss, New York, pp 77 – 97

33. Nohl H and Hegner D (1978) Do mitochondria produce oxygen radicals in vivo? Eur J Biochem 82: 563 – 567
34. Nohl H, Jordan W (1984) The biochemical role of ubiquinone and ubiquinone-derivatives in the generation of hydroxyl-radicals from hydrogen-peroxide. In: Bors W, Saran M and Tait D (eds) Oxygen radicals in chemistry and biology. de Gruyter, Berlin, pp 155 – 163
35. Nohl H, Jordan W, Hegner D (1982) Mitochondrial formation of OH•-radicals by a ubisemiquinone-dependent reaction. An alternative pathway of the iron-catalysed Haber-Weiss-cycle. Hoppe Seyler's Z Physiol Chem 363: 599 – 607
36. Norling B, Glazek E, Nelson NB, Ernster L (1974) Studies with ubiquinone-depleted submitochondrial particles; quantitative incorporation of small amounts of ubiquinone and the effects on the NADH- and succinate-oxidase activities. Eur J Biochem 47: 475 – 482
37. Ono T, Cutler RG (1978) Age-dependent relaxation of gene repression: increase of endogenous murine leukemia virus-related and globin-related RNA in brain and liver of mice. Proc Natl Acad Sci USA 75: 4431 – 4435
38. Perry TL, Godin DV, Hansen S (1982) Parkinson's disease: a disorder due to nigral glutathione deficiency? Neurosci Lett 33: 305 – 310
39. Piko L, Bulpitt KJ, Meyer R (1984) Structural and replicative forms of mitochondrial DNA in tissues from adult and senescent BALB/c mice and Fisher 344 rats. Mech Ageing Dev 26: 113 – 131
40. Rehncrona S, Folbergrová J, Smith DS, Siesjö BK (1980) Influence of complete and pronounced incomplete cerebral ischemia and subsequent recirculation on cortical concentrations of oxidized and reduced glutathione in the rat. J Neurochem 34: 477 – 486
41. Rich PR, Bendall DS (1980) The kinetics and thermodynamics of the reduction of cytochrome *c* by substituted *p*-benzoquinols in solution. Biochim Biophys Acta 592: 506 – 518
42. Sohal RS, Allen RG (1986) Relationship between oxygen metabolism, aging and development. Adv Free Radical Biol Med 2: 117 – 160
43. Trumpower BL (1978) Restoration of ubiquinone-pool function, tight behaviour to succinate dehydrogenase, having latent reconstructive activity of ubiquinone-cytochrome *c* reductase complex. Biochem Biophys Res Commun 83: 528 – 535
44. Trumpower BL, Simmons Z (1979) Diminished inhibition of mitochondrial electron transfer from succinate to cytochrome *c* by thenoyltrifluoroacetone induced by antimycin. J Biol Chem 254: 4608 – 4616
45. Tzagoloff A, Macino G, Sebald W (1979) Mitochondrial genes and translation products. Annu Rev Biochem 48: 419 – 441
46. Von Hahn HP, Heim JM, Eichhorn GL (1970) The effect of divalent ions on the isolation of proteins from rat liver nucleoprotein. Biochim Biophys Acta 214: 509 – 519
47. Wagner AP, Psarrou E, Wagner LP (1982) Age changes of the isoelectric points of nonhistone chromosomal proteins from rat liver in the pH range 5 to 8. Exp Gerontol 17: 359 – 364
48. Winter DB, Bruyninckx WJ, Foulke FG, Grinich NP, Mason HS (1980) Location of heme a on subunits I and II and copper on subunit II of cytochrome *c* oxidase. J Biol Chem 255: 11408 – 11414
49. Yong VW, Perry TL, Krisman AA (1986) Depletion of glutathione in brainstem of mice caused by *N*-methyl-4-phenyl-1,2,3,6-tetrahydropyridine is prevented by antioxidant pretreatment. Neurosci Lett 63: 56 – 60
50. Yu L, Yu CA, King TE (1978) The indispensability of phospholipid and ubiquinone in mitochondrial electron transfer from succinate to cytochrome *c*. J Biol Chem 253: 2657 – 2663
51. Zegarelli-Schmidt RC, Goodman R (1981) The dipteran as a model system in cell and molecular biology. Int Rev Cytol 71: 245 – 363

Oxidative, Energy and Related Brain Metabolism During Normal Aging

S. Hoyer

Introduction

Increasing life expectancy has raised health problems with respect to "normal" aging and particularly to age-related disorders. This holds true for several brain diseases in middle and old age, such as stroke (brain ischemia), Parkinson's disease, and dementia of either vascular or Alzheimer type.

There is ample evidence that normal brain function is coupled to an undisturbed circulation and substrate supply. Any change in cerebral blood flow and/or in the composition of brain nutrients will alter the oxidative and energy metabolism in the brain and thus give rise to the beginning of brain disorders of different kinds. Otherwise, autochthonously occurring neurodegeneration in, e.g., dementia of Alzheimer type, may be associated with severe changes in cerebral oxidative and energy metabolism [17]. In this short review, changes in oxidative, energy, and related brain metabolism during normal aging will be elucidated. It is concluded that age may be considered a risk factor for abnormal brain function.

With respect to oxidative brain metabolism, the mature, healthy, nonstarved mammalian brain uses glucose only as a source of energy in the form of ATP to meet its functional and structural requirements, such as neurotransmission, synaptic function, cellular calcium homeostasis, axoplasmatic flux, and cellular integrity [3, 7, 8, 13, 28, 36, 37]. Thus glucose metabolism in the brain plays a pivotal role in maintaining cellular homeostasis. Perturbations in this metabolic pathway therefore may generate cellular dysfunction leading to neuropsychiatric deficits.

Brain Glucose and Related Metabolism

The Normal Adult Brain

The control of cerebral glucose metabolism is based upon various mechanisms. Glucose uptake from arterial blood into the brain across the blood-brain barrier by means of a carrier-mediated transport mechanism is obvioulsy insulin-dependent [1, 12, 20]. The glycolytic breakdown of glucose is controlled by the enzymes hexokinase and pyruvate kinase, working in concert under the predominance of phosphofructokinase. Oxidation of the gly-

Reihe der Villa Vigoni, Band 1
Biology of Aging
Zwilling and Balduini (Eds.)

colytically formed pyruvate is started by means of the multienzyme complex pyruvate dehydrogenase (PDH). Finally, the energy state of the brain tissue influences cellular glucose metabolism via feedback mechanisms [36].

The mitochondrial PDH is of paramount importance in glucose metabolism. It provides the acetyl groups for subsequent oxidation and energy production on the one hand and for acetylcholine synthesis on the other. The activity of the enzyme choline acetyltransferase, which catalyzes acetylcholine formation, is closely linked functionally to PDH and is stimulated by insulin [23, 33].

The activation state of PDH depends on its state of phosphorylation. Phosphorylation inhibits and dephosphorylation activates this multienzyme complex [26]. The activity of the dephosphorylated (active) form of PDH after membrane depolarization is increased by cytosolic Ca^{2+}, more extensively in brain mitochondria than in synaptosomes. When fuelled by pyruvate, calcium accumulation by mitochondria was found to be closely linked to PDH activity. Phosphorylation of the PDH α-subunit may interfere with the calcium-buffering capacity of mitochondria, and PDH activity was markedly reduced when mitochondrial Ca^{2+} uptake was inhibited. Thus, the activation state of PDH contributes to cellular Ca^{2+} homeostasis [10]. On the other hand, activation of PDH decreased in proportion to the decline in pyruvate-supported calcium transport into hippocampal mitochondria following the induction of entorhinal cortex lesions. Independently of the mitochondrial Ca^{2+} concentration, the amount of the active, dephosphorylated form of PDH was increased by stimulation with insulin, and it changes in the opposite direction to the concentration of adenine nucleotides [19].

PDH activity is inhibited by its product, acetyl CoA, for example when fatty acids or ketone bodies are used for oxidation. Under normal conditions, glucose carbon is rapidly transferred into amino acids via the tricarboxylic acid cycle and the γ-aminobutyric acid (GABA) shunt. Glutamate, glutamine, aspartate, and GABA are formed in the greatest abundance [43]. At least two compartments of these acidic amino acids, one of which is a storage compartment, are postulated in the brain. These glucoplastic amino acids may serve in part as a fuel reserve for emergency conditions when glucose is lacking. Moreover, glutamate and aspartate function as excitatory neurotransmitters active throughout nearly the entire brain but particularly in the entorhinal afferents and the Schaffer collaterals, both of which end in the CA 1 subfield of the hippocampus and also in its mossy fibers. Cholinergic neurons receive glutamatergic afferents [5]. Following K^{+}-evoked depolarization, aspartate, glutamate, and GABA were preferentially released from neocortical tissue, while glutamate efflux was calcium-dependent. After its release from nerve endings, glutamate binds with a high affinity to postsynaptic dendritic membranes [34]. Furthermore, glutamate induced the release of GABA from cerebral cortex interneurons and enhanced GABA-activated conductance in hippocampal pyramidal neurons at a concentration below that required for excitation. Hence, it is obvious that the metabolism of glucose and related compounds in the brain may play a central role in maintaining normal neuronal function. Disturbances in glucose metabolism may give rise to neuronal damage and cause abnormal mental states, varying in severity with the degree of cellular damage.

The Aging Brain

Several studies show that mental capacity is closely associated with the brain perfusion activity necessary for both substrate supply and waste product transportation and also with oxidative and energy metabolism. Whether or not cerebral blood flow and metabolism decrease with advancing age is a debated issue. From a review of the literature, it may be deduced that there is obviously only a minor reduction in cerebral blood flow and in oxygen and glucose utilization rates between the 3rd – 4th and the 7th – 8th decades of life, with a more severe decrease in later life [15]. Thus, a threshold phenomenon concerning brain perfusion and nutrition may be assumed with advancing age. The same holds true for phosphofructokinase [30], choline acetyltransferase [19], and acetylcholine esterase activities and for acetylcholine synthesis [2, 4] in the normally aging human brain. Furthermore, the activities of the cAMP-dependent protein kinase and glutamate decarboxylase are markedly diminished after the age of 70 years, whereas no age-related variations are seen in Na^+/K^+-ATPase and Mg^{2+}-ATPase, which break down ATP [2]. Reductions in the above brain parameters before this threshold of 70 years in human life may therefore be disease-oriented rather than age-related (29, 38, 40).

On the basis of animal studies, it may be deduced that glucose consumption in the brain cortex and in some subcortical nuclei diminishes from the developmental phase to adulthood but then does not decrease further with progression to old age.

At the cellular level, a moderate but steady decline in glycolytic compounds concentrations was found from development to senescence; however, the diminution was not evenly distributed over the different periods of life. Glucose, fructose-1,6-diphosphate, and ATP levels decreased mostly from the developmental stage to adulthood, whereas those of pyruvate and creatine phosphate fell mostly from adulthood to senescence under resting conditions [14].

The changes in amount of glycolytic compounds were associated with age-related reductions in the activities of the enzymes controlling glycolytic flux, hexokinase and phosphofructokinase [18, 24]. On the other hand, only slight variations were found in the oxidation processes of the tricarboxylic acid cycle and respiratory chain with aging. The decrease in malate concentration found in senescent rats [14] may be consistent with reduced activity of malate dehydrogenase, indicating minimally diminished tricarboxylic acid cycle activity. No age-related changes were reported in the enzyme activities of PDH (active and total), citrate synthase, and oxidized nicotinamide-adenine dinucleotide (NAD^+)-isocitrate dehydrogenase , fumarase, and NAD^+-malate dehydrogenase [6, 25], although observations concerning NAD^+-isocitrate dehydrogenase are inconsistent. Studies on cytochrome a, a_3, as the final member of substrate oxidation that can react directly with molecular oxygen, revealed no age-related changes [39], although oxygen consumption in the rat cerebral cortex was found to decrease gradually with age [32], and $^{14}CO_2$ production diminished by 30 % in senescence [9]. The neurotransmitters that derive from glucose breakdown in the brain also show reduced concentrations with aging. Acetylcholine synthesis showed a 65 % decline in senescence as compared with young adulthood. On the other hand, acetylcholine release was reduced by 25 % in the same study [9], and by 50 % in another [31]. The number of muscarinic receptors in the dorsal hippocampus fell by 22 % in aged rats [27]. On the whole, these findings are indicative of the reduced capacity of the acetylcholinergic system in old age.

Glutamate and aspartate are diminished in the aged brain. However, whether glutamate synthesis from glucose stays unchanged or is reduced remains a subject of controversy, since conversion of glucose carbon into glutamate, glutamine, GABA, aspartate, and alanine decreases in the cerebrum with aging [41]. As mentioned above, the basal release of glutamate in the striatum and hippocampus increased with aging, so its extracellular concentration was raised: after K^+ stimulation this level remained unaffected, and glutamate binding to hippocampal membranes increased progressively because the number of binding sites and not the receptor affinity for glutamate had increased (for more details, see chapter by W.E. Müller et al., this volume).

Additional aspects related to cerebral glucose metabolism should be considered: (a) the activity of free radicals, which are potent in cellular destruction via proteolysis and lipid peroxidation (for more details, see chapter by G. Benzi and D. Curti, this volume). The formation of free radicals increases with advancing age [11]. Tissue lactacidosis dramatically enhances the formation of free radicals in brain tissue, by mediating an increased dissociation of catalytic iron from proteins of the transferrin type. (b) In the aged brain, a loss of nerve growth factor sensitivity, i.e., a reduced expression of nerve growth factor (NGF) receptors, has been found [22]. Since NGF is known as an important compound in cellular repair after damage [35], the reduced expression of its receptor may point to a diminished repair capacity with age. (c) A delayed lysosomal response with advanced age was reported, which may contribute to an impairment in the elimination of cellular degeneration debris and thus damage cellular function further [42].

The discussion of age-related changes in glucose and related metabolism and other metabolic pathways does not explain why some biological parameters, which are considered important in maintaining normal neuronal function, become abnormal after the 7th or 8th decade of human life or late in life in rodents, while others remain unchanged. The majority of the biological abnormalities in the nondiseased aging process of the brain may indicate a first slight incipient perturbation in neuronal homeostasis [21]. This includes reductions in neuronal glucose utilization due to a decreased glycolytic flux, slight aberrations in glucose oxidation, diminished acetylcholine synthesis and release, and increases in cytosolic Ca^{2+} concentration, extracellular glutamate, and glutamate binding, as well as free radical formation enhancement under lactacidosis. These metabolic variations in the "milieu intérieur" in all probability characterize cellular stress, which subsequently may give rise to the morphological destruction of highly vulnerable neurons during aging, as discussed above. Concomitantly, repair and cleaning systems of the cell are diminished in function. Thus, age may be considered to be a risk factor for the generation of abnormal brain function or even of long-lasting brain disorders subsequent to stressful events. Ischemia to the brain may serve as an example [16].

Conclusion

The brain aging process per se is accompanied by metabolic aberrations demonstrating an incipient perturbation of neuronal homeostasis, which includes neuronal glucose metabolism, acetylcholine synthesis and release, extracellular concentration and receptor binding of glutamate, cytosolic Ca^{2+} homeostasis, free radical formation, and cellular repair and cleaning processes. It seems likely (and there is some support of this assumption) that the sum of

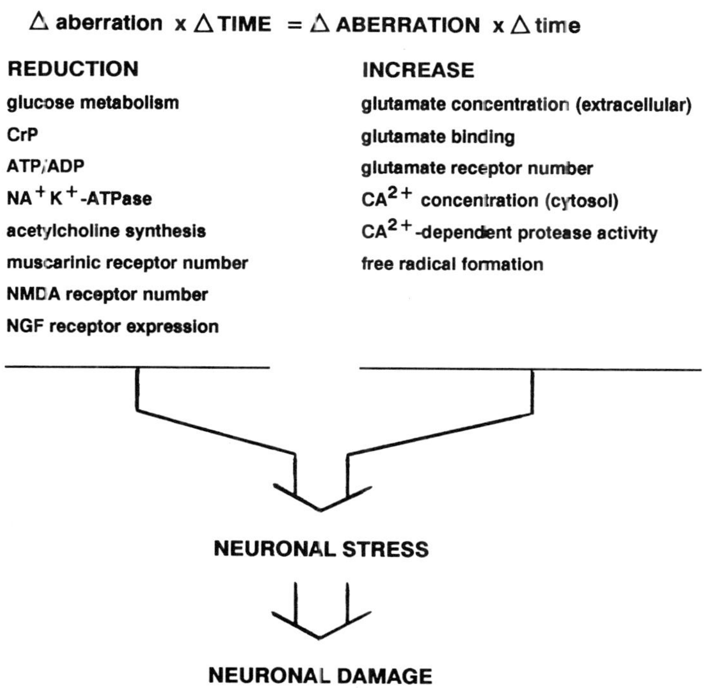

Fig. 1. Summary of abnormalities leading to neuronal damage. *NGF*, nerve growth factor; *Crp*, creatine phosphate; NMDA, *N*-methyl-D-aspartate

these cellular impairments may develop in a self-propagating manner and that this process could lead to destruction of highly vulnerable neurons, particularly in the brain cortex and hippocampus. One of these impairments on its own would not have the capacity to destroy cells and can obvioulsy be repaired. However, the sum of numerous small abnormalities is able to damage and destroy cells (Fig. 1).

References

1. Bachelard HS (1971) Specific and kinetic properties of monosaccharide uptake into guinea pig cerebral cortex in vitro. J Neurochem 13: 213 – 222
2. Bowen DM, Smith CB, White P, Flack RHA, Carrasco LH, Gedye JL, Davison AN (1977) Chemical pathology of the organic dementias: II. Quantitative estimation of cellular changes in postmortem brain. Brain 100: 427 – 453
3. Cohen PJ, Alexander SC, Smith TC, Reivich M, Wollmann H (1967) Effects of hypoxia and normocarbia on cerebral blood flow and metabolism in conscious man. J Appl Physiol 23: 183 – 189
4. Davies P (1979) Neurotransmitter-related enzymes in senile dementia of the Alzheimer type. Brain Res 171: 319 – 327
5. Davies SW, McBean GJ, Roberts PJ (1984) A glutamatergic innervation of the nucleus basalis/substantia innominata. Neurosci Lett 45: 105 – 110
6. Deshmukh DR, Owen OE, Patel MS (1980) Effect of aging on the metabolism of pyruvate and 3-hydroxybutyrate in nonsynaptic and synaptic mitochondria from rat brain. J Neurochem. 34: 1219 – 1224
7. Erecinska M, Silver IA (1989) ATP and brain function. J Cereb Blood Flow Metab 9: 2 – 19
8. Gibbs EL, Lennox WG, Nims LF, Gibbs FA (1942) Arterial and cerebral venous blood. Arterial-venous differences in man. J Biol Chem 144: 325 – 332
9. Gibson GE, Peterson C (1981) Aging decreases oxidative metabolism and the release and synthesis of acetylcholine. J Neurochem 37: 978 – 984
10. Hansford RG, Castro F (1985) Role of Ca^{2+} in pyruvate dehydrogenase interconversion in brain mitochondria and synaptosomes. Biochem J 227: 129 – 136
11. Harman D (1981) The aging process. Proc Natl Acad Sci USA 78: 7124 – 7128
12. Hertz MM, Paulson OB, Barry DI, Christiansen JS, Svendsen PA (1981) Insulin increases glucose transfer across the blood-brain barrier. J Clin Invest 67: 597 – 604
13. Hoyer S (1970) Der Aminosäurenstoffwechsel des normalen menschlichen Gehirns. Klin Wochenschr 48: 1239 – 1243
14. Hoyer S (1985) The effect of age on glucose and energy metabolism in brain cortex of rats. Arch Gerontol Geriatr 4: 193 – 203
15. Hoyer S (1986) Senile dementia and Alzheimer's disease. Brain blood flow and metabolism. Prog Neuropsychopharmacol Biol Psychiatry 10: 447 – 478
16. Hoyer S, Krier C (1986) Ischemia and the aging brain. Studies on glucose and energy metabolism in rat cerebral cortex. Neurobiol Aging 7: 23 – 29
17. Hoyer S, Oesterreich K, Wagner O (1988) Glucose metabolism as the site of the primary abnormality in early-onset dementia of Alzheimer type? J Neurol 235: 143 – 148
18. Iwangoff P, Armbruster R, Enz A, Meier-Ruge W, Sandoz P (1980) Glycolytic enzymes from human autoptic brain cortex: normally aged and demented cases. In: Roberts PJ (ed) Biochemistry of dementia. Wiley, Chichester, pp 258 – 262
19. Jope R, Blass JP (1976) The regulation of pyruvate dehydrogenase in brain in vivo. J Neurochem 26: 709 – 714
20. Kahn CR (1985) The molecular mechanism of insulin action. Ann Rev Med 36: 429 – 451
21. Khachaturian ZS (1990) The role of calcium regulation in brain aging: reexamination of a hypothesis. Aging 1: 17 – 34

22. Koh S, Loy R (1988) Age-related loss of nerve growth factor sensitivity in rat basal forebrain neurons. Brain Res 440: 396 – 401
23. Kyriakis JM, Hausmann RE, Peterson SW (1987) Insulin stimulates choline acetyltransferase activity in cultured embryonic chicken retina neurons. Proc Natl Acad Sci USA 84: 7463 – 7467
24. Leong SF, Clark JB (1984) Regional enzyme development in rat brain. Enzymes associated with glucose utilization. Biochem J 218: 131 – 138
25. Leong SF, Lim JCK, Clark JB (1981) Energy-metabolizing enzymes in brain regions of adult and aging rats. J Neurochem 37: 1548 – 1556
26. Linn TC, Pettit FH, Reed LJ (1969) Alpha-keto acid dehydrogenase complexes: X. Regulation of the activity of the pyruvate dehydrogenase complex from beef kidney mitochondria by phosphorylation and dephosphorylation Proc Natl Acad Sci USA 62: 234 – 241
27. Lippa AS, Critchett DJ, Ehlert F, Yamamura HI, Enna SJ, Bartus RT (1981) Age-related alterations in neurotransmitter receptors: an electrophysiological and biochemical analysis. Neurobiol Aging 2: 3 – 8
28. Lipton P, Whittingham TS (1982) Reduced ATP concentration as a basis for synaptic transmission failure during hypoxia in the in vitro guinea pig hippocampus. J Physiol 325: 51 – 65
29. London ED, Nespor SM, Ohata M, Rapoport SI (1981) Local cerebral glucose utilization during development and aging in the Fischer-344 rat. J Neurochem 37: 217 – 221
30. Meier-Ruge W, Hunziker O, Iwangoff P, Reichlmeier L, Schultz U (1980) Effect of age on morphological and biochemical parameters of the human brain. In: Stein DC (ed) The psychobiology of aging: problems and perspectives. Elsevier, North Holland, pp 297 – 317
31. Pedata F, Slavikova J, Kotas A, Pepeu G (1983) Acetylcholine release from rat cortical slices during postnatal development and aging. Neurobiol Aging 4: 31 – 35
32. Peng MT, Peng YI, Chen FN (1977) Age-dependent changes in the oxygen consumption of the cerebral cortex, hypothalamus, hippocampus and amygdaloid in rats. J Gerontol 32: 517 – 522
33. Perry EK, Perry RH, Tomlinson BE, Blessed G, Gibson PH (1980) Coenzyme A acetylating enzymes in Alzheimer's disease: possible cholinergic "compartment" of pyruvate dehydrogenase. Neurosci Lett 18: 105 – 110
34. Rothman SM, Olney JW (1986) Glutamate and the pathophysiology of hypoxic-ischemic brain damage. Ann Neurol 19: 105 – 111
35. Shelton DL, Reichardt LF (1986) Studies on the expression of the nerve growth factor (NGF) gene in the central nervous system: level and regional distribution of NGF mRNA suggest that NGF functions is a trophic factor for several distinct populations of neurons. Proc Natl Acad Sci USA 83: 2714 – 2718
36. Siesjö BK (1978) Brain energy metabolism. Wiley, Chichester, pp 1 – 28, 151 – 209
37. Siesjö BK (1981) Cell damage in the brain: a speculative synthesis. J Cereb Blood Flow Metab 1: 155 – 185
38. Smith CB, Goochee C, Rapoport I, Sokoloff L (1980) Effects of ageing on local rates of cerebral glucose utilization in the rat. Brain 103: 351 – 365
39. Sylvia AL, Rosenthal M (1978) The effect of age and lung pathology on cytochrome a, a_3 redox levels in rat cerebral cortex. Brain Res 146: 109 – 112
40. Takei J, Fredericks WR, London ED, Rapoport SI (1983) Cerebral blood flow and oxidative metabilism in conscious Fischer-344 rats of different ages. J Neurochem 40: 801 – 805
41. Tyce GM, Wong KL (1980) Conversion of glucose to neurotransmitter amino acids in the brain of young and aging rats. Exp Gerontol 15: 527 – 532
42. Vijayan VK, Cotman CW (1983) Lysosomal enzyme changes in young and aged control and entorhinal-lesioned rats. Neurobiol Aging 4: 13 – 23
43. Wong KL, Tyce GM (1983) Glucose and amino acid metabolism in rat brain during sustained hypoglycemia. Neurochem Res 8: 401 – 415

Different Susceptibilities of Protein Thiol Groups to Oxidative Modifications: Implications in Aging Research

G. Bellomo, F. Mirabelli, R. Fulceri, and A. Benedetti

Introduction

Aging is a biological process that is believed to originate from multicomponent and multifactorial mechanisms affecting the animal organism as a whole. Several experimental models have been developed in a variety of animals to investigate the biology and pathophysiology of aging. Studies performed with them have alternatively provided evidence supporting a wide variety of different pathogenetic mechanisms. Recently, the so-called oxidative theory of aging has obtained increasing experimental support (for a series of reviews, see [5]). According to this theory, the increased generation of oxygen free radicals and oxygen reactive species together with the decline in the antioxidant defenses generate an oxidative stress and a progressive impairment of the physiological activities. This view is supported by the demonstration of a decreased activity of some antioxidant enzymes in elderly animals, together with the appearance of direct or indirect signs of oxidative damage (such as lipid peroxidation).

In the past few years, cell cultures and isolated cell systems have been extensively employed to investigate oxidative cell injury. Although these simple cellular models cannot be used to mimic what exactly occurs in a whole animal or organism, they can be extremely useful in investigating some of the molecular mechanisms and in detecting some of the specific targets of oxidative cellular insults.

Our laboratory has been actively engaged in studying the metabolic and structural alterations occurring in a variety of mammalian cells exposed to oxygen free radicals and reactive species. Here we report some of our recent findings on the occurrence of thiol modifications during oxidative stress, on the different susceptibility of these functional groups to oxidation, and on the role of this process in triggering irreversible cellular injury.

Disruption of Intracellular Thiol Homeostasis Following Exposure to Oxygen Reactive Species

One of the most important cellular first-line defenses against oxygen free radicals is definitely represented by glutathione (in the reduced form, GSH) and by several glutathione-related enzymes [4]. In addition to its well-charac-

Reihe der Villa Vigoni, Band 1
Biology of Aging
Zwilling and Balduini (Eds.)

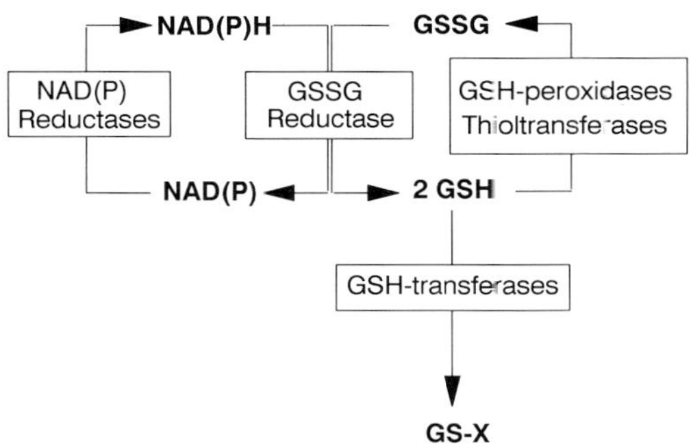

Fig. 1. Schematic representation of the glutathione redox cycle. *NAD (P)*, nicotinamide adenine dinucleotide (phosphate); *GSH*, reduced glutathione; *GSSG*, glutathione disulfide

terized function as an electrophile binding directly or indirectly (through the activity of glutathione-S transferases) several nucleophilic reactive intermediates, GSH may act as a reducing agent in the metabolism of hydrogen peroxide, organic hydroperoxide, and phospholipid hydroperoxide. These reactions are catalyzed by a number of glutathione peroxidases located in different cellular compartments such as the cytoplasm, nucleus, mitochondrion and bound to cellular membranes [32]. Glutathione disulfide (GSSG) is formed during this reaction and under physiological conditions is reduced to GSH by the enzyme glutathione reductase at the expense of reduced nicotinamide-adenine dinucleotide phosphate [NAD(P)H] that is oxidized to $NAD(P)^+$ (Fig. 1). GSSG can also be formed through several other metabolic routes, such as during the activity of the enzyme thiol-disulfide oxidoreductase that catalyzes thiol-disulfide exchange at the protein level or as a result of the reactivity of the thiyl radical formed by a direct attack of oxygen reactive species on GSH [18]. However, when the rate of GSSG formation overwhelms that of its reduction, or when the activity of glutathione reductase is decreased, GSSG accumulates in the cytosol and is subsequently extruded from the cell by a specific ATPase coupled to a transport system. This may lead to a substantial decrease of cellular GSH concentration and may sensitize the cell toward oxidative injury [11].

Studies performed both in vitro and in vivo have demonstrated that GSH depletion below a critical threshold triggers a series of events that ultimately leads to irreversible cell injury and death [11]. They include the initiation and propagation of lipid peroxidation and the oxidation and depletion of sulfhydryl groups in proteins. Little is known about the mechanisms underlying lipid peroxidation in GSH-depleted cells. Investigators in this field have alternatively proposed that GSH depletion unmasks a sort of mild oxidative stress physiologically occurring in normal cells or may relocate a pool of iron from intracellular stores, thus making it available for catalyzing the initial steps of lipid peroxidation.

The other pathophysiologically relevant event occurring in the GSH-depleted status is the oxidation and/or depletion of protein thiol groups. Several mechanisms participating in this process have been identified. They include direct oxidation by unbuffered oxygen reactive species [13], formation of mixed disulfides with GSSG [3] or other disulfides, oxidative cross-linking via disulfide bonds [21], and direct alkylation by reactive intermediates or free radicals [30]. Recently, two different investigations discovered more complex mechanisms.

In a study performed on isolated hepatocytes treated with paracetamol, it was demonstrated that direct alkylation of critical sulfhydryl groups in proteins by reactive paracetamol metabolites leads to a sustained increase in cytosolic free Ca^{2+} concentration which, via a Ca^{2+}-dependent protease, activates the enzyme xanthine oxidase [31]. This activation generates a substantial amount of superoxide anion free radical and other oxygen reactive species that, in turn, oxidize other protein sulfhydryl groups. It thus appears that the two main mechanisms of protein thiol depletion, namely oxidation and alkylation, may be linked together.

Moreover, investigations performed in vivo during allyl alcohol toxicity by Pompella and coworkers [26] have revealed that, in the same cell and in the same experimental model, protein thiol depletion may occur by two different mechanisms, depending on the cellular compartment involved. Thiol depletion in the cytoplasm is primarily due to alkylation by the reactive allyl alcohol metabolite acrolein that is, indeed, formed in this compartment. However, depletion of sulfhydryl groups of membrane proteins appears directly linked to the binding of aldehydic products generated by lipid peroxidation, a process occurring during allyl alcohol metabolism but limited to cellular membranes.

The demonstration that protein thiol oxidation and/or depletion occurs in oxidative stress-challenged cells led to the proposal that this could be an important step in the development of irreversible cell injury. This hypothesis was essentially based on the demonstration that manipulation of cellular thiol status by GSH depletion or, on the contrary, by thiol reducing agents, was extremely efficient in modulating the cytotoxicity of oxygen reactive species [2]. In addition, it appeared that the relationship between thiol oxidation and cytotoxicity was not exclusively quantitative but, in some experimental models, was linked to the impairment of an extremely limited pool of protein sulfhydryl groups [23]. This led to the search for individual proteins or set of proteins absolutely essential for cell survival and whose activity could be affected by oxidation of critical sulfhydryl residues. Table 1 lists some of the proteins that have been identified so far. As a consequence of oxidative inactivation of these proteins, the permeability barrier of the plasma membrane is lost [17], the intercellular communication can be dramatically impaired [4], the efficiency of energy production is reduced [8], the intracellular energy stores are rapidly exhausted, intracellular ion homeostasis is disrupted [24], and the cellular metabolic and structural organization is markedly altered [25].

Table 1. Identification of cellular proteins whose sulfhydryl groups are affected during oxidative stress generated by oxygen free radicals

Protein	Consequences
Plasma membrane proteins	Increased membrane permeability to small molecules and ions
Mitochondrial electron transport enzymes and inner membrane proteins	Impaired mitochondrial respiration, inhibition of oxidative phosphorylation
Glyceraldehyde-3-P-dehydrogenase and other glycolytic enzymes	Inhibition of glycolysis, impaired ATP production
Ion-motive ATPases	Disruption of intracellular ion homeostasis
Surface receptors, G proteins, and protein kinases	Impaired signal transduction
Cytoskeletal proteins (microfilaments, microtubules, intermediate filaments)	Alterations in cell shape and organization

Alterations of Cytoskeleton Structure and Organization During Oxidative Stress

In the past few years we have actively investigated the possibility that cytoskeletal alterations may play a role in the development of irreversible injury during oxidative stress in a variety of different cells. As the experimental model we employed the metabolism of redox cycling quinones (such as 2-methyl-1,4-naphtoquinone, menadione) that are reduced to semiquinone free radicals by one-electron reduction catalyzed by some mitochondrial and endoplasmic reticular flavoenzymes. In the presence of molecular oxygen, the semiquinone radicals are reoxidized to the parent quinones with the contemporary generation of superoxide anion free radical and other oxygen reactive species [29].

Cytoskeletal proteins exist in three classes of fibers, microfilaments, microtubules, and intermediate filaments, that differ markedly in terms of their protein composition and functions, as summarized in Table 2 (see also [12] for a review). The cytoskeleton controls several aspects of cell membrane physiology, including the compartmentation of enzymes and metabolic processes, the mobility of surface receptors for hormones and growth factors, the transport of cellular organelles, and the migration and attachment of the whole cell. Evidence suggests that the cytoskeleton is involved in oxidative cell injury. In oxidative stress-challenged cells, receptor internalization and recycling are impaired, exocytosis and secretion are inhibited, adherent cells detach from the growing substrate, and cell morphology is modified [5]. The appearance of numerous small surface protrusions is, in fact, a characteristic feature of oxidatively injured cells, a morphologic alteration indistinguishable from that occurring in cells exposed to cytoskeletal toxins such as cytochalasin B and phalloidin [19].

Table 2. Organization of cytoskeletal proteins in classes of fibers in non muscle cells

Classes	Composition	Fiber diameter	Functions
Microfilaments	Actin Actin-binding proteins	7 nm	Cell locomotion Control of cell shape; Endo- and exocytosis
Microtubules	Tubulin MAPs	24 nm	Axonal transport Organelle transport Formation of mitotic spindle Motion of cilia and flagella
Intermediate filaments	Keratins Desmin Vimentin NP GFAP	7 – 15 nm	Control of cell shape Nucleus-membrane interaction Subcellular organization of the cytoplasm

NF, neurofilament polypeptides; GFAP, glial fibrillar acidic protein(s); MAPs, microtubule-associated proteins

The incubation of cultured cells with a cytotoxic concentration of the oxidant menadione causes a marked depletion of total cellular protein sulfhydryl groups [13]. The isolation of the different cytoskeletal components from treated cells revealed, in fact, that microfilaments, microtubules, and intermediate filaments are almost equally affected by the process of protein thiol depletion (Table 3) [6]. Moreover, to investigate the possibility that oxidative stress-induced depletion of cytoskeletal protein sulfhydryl groups could be a general phenomenon, four different cell types were exposed to menadione, and the various cytoskeletal fractions were extracted and analyzed. As shown in Table 3, all cell lines investigated exhibited a similar decrease in protein thiols that affected all the cytoskeletal fractions investigated.

Table 3. Depletion of cytoskeletal protein sulfhydryl groups during the metabolism of 2-methyl-1,4-naphthoquinone in various cell lines

Cell line	Protein sulfhydryl groups (% of control)		
	Whole cytoskeleton	Intermediate filaments fraction	Microtubule fraction
Hepatocytes (primary cultures)	31	45	38
McCoy's cells	36	51	44
BT-20 cells	42	53	43
HeLa cells	38	44	48

The cell types were incubated for 1 h in the absence or in the presence of 200 μM menadione. The indicated cytoskeletal fractions were isolated and protein sulfhydryl groups were measured. For experimental details, see Bellomo et al. [6]

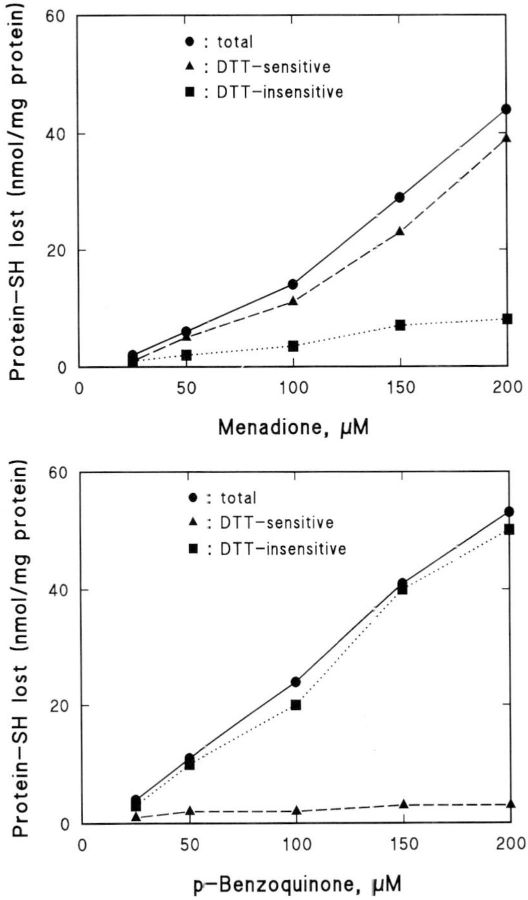

Fig. 2. Different mechanisms of thiol depletion induced in cytoskeletal proteins by the metabolism of menadione and *p*-benzoquinone. Isolated hepatocytes were incubated for 60 min with the indicated quinone concentration and subsequently treated for an additional 30 min with or without 5 μM dithiothreitol (DTT). The cytoskeletal fraction was then extracted and the amount of protein thiols quantitated as described by Mirabelli et al. [21]

A more detailed biochemical analysis of the depletion of cytoskeletal protein sulfhydryl groups has led to the characterization of several different processes. These investigations have been performed using two different quinones, namely menadione, which causes primarily thiol oxidation, and *p*-benzoquinone, that causes thiol alkylation [30]. As shown in Fig. 2, the two experimental conditions can be, in fact, differentiated on the basis of the reversibility of thiol depletion upon treatment with the thiol reducing agent dithiothreitol (DTT): most of the -SH residues lost during the metabolism of menadione are sensitive to DTT, and in contrast, most of those lost during the metabolism of *p*-benzoquinone are completely insensitive to DTT.

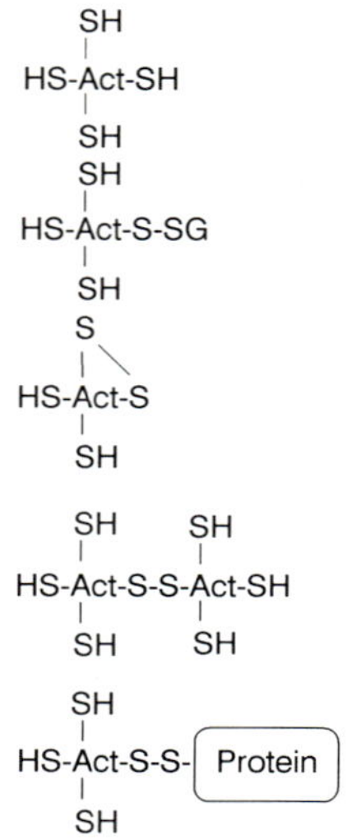

Fig. 3. Potential mechanisms involved in the oxidation of thiol groups in cytoskeletal actin

Three different mechanisms can be theoretically involved in thiol oxidation reversed by DTT: (a) mixed disulfide formation, (b) intramolecular and (c) intermolecular disulfide bond formation. Studies, specifically performed on cytoskeletal actin, have indeed revealed that its four -SH residues can be targets for these processes (Fig. 3). The formation of mixed disulfides between actin and GSSG and of an intramolecular disulfide bond results in the modification of the surface charge of the actin molecule. This can be easily detected with polyacrylamide gel electrophoresis analysis under non reducing conditions as an increased mobility of the molecule itself, generating the so-called fast-migrating actin [30]. It is interesting to notice that similar modifications have been detected in several cytoskeletal proteins of aged erythrocytes [1].

Intermolecular disulfide bonds between different actin molecules or between actin and other proteins have been reported in the cytoskeleton of quinone-treated hepatocytes [21]. This oxidative protein cross-linking leads to the formation of large molecular weight aggregates that have been detected both biochemically and morphologically in challenged cells [6]. In addition to actin, the other proteins included in these aggregates may be cytoskeletal (as occurs in platelets with α-actinin and other actin-binding proteins) [22] or noncytoskeletal in origin. This is clearly indicated by data reported in Fig. 4 and refers to the ability of DTT to release noncytoskeletal proteins from the cytoskeleton extracted from hepatocytes exposed to the oxidizing quinone menadione or to the alkylating quinone *p*-benzoquinone.

Little is known, however, of the pathophysiological consequences of the oxidation of sulfhydryl groups in cytoskeletal proteins. It has been hypothesized that the formation of oxidative cross-linked aggregates including microfila-

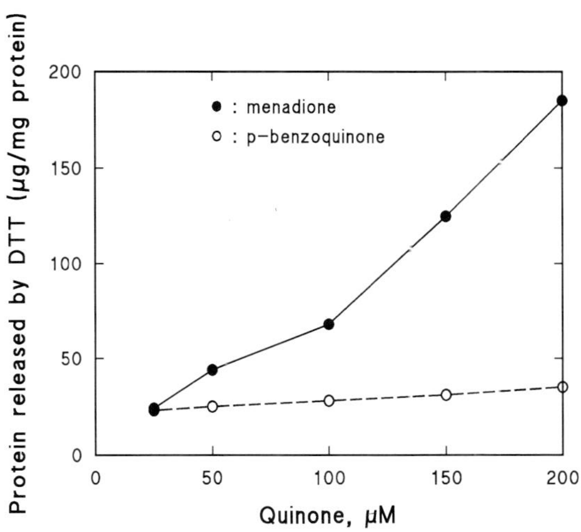

Fig. 4. Formation of intermolecular disulfide bridges sensitive to dithiothreitol in the cytoskeleton of hepatocytes during the metabolism of redox cycling and alkylating quinones. Isolated hepatocytes were incubated for 60 min with the indicated quinone concentration. The cytoskeletal fraction was then extracted and subsequently incubated for an additional 30 min with or without 5 m*M* dithiothreitol. The amount of protein released from the cytoskeleton was then quantitated. For experimental details, see Thor et al. [30]

mentous actin and actin-binding proteins would lead to the appearance of discrete regions in the plasma membrane not anchored to the cytoskeletal network [7]. These regions would represent sites of weakness where protrusions (blebs) could be formed and progress until the rupture of the surrounding membrane, thus precipitating cell death. In fact, the addition of thiol reducing agents to menadione-treated cells causes the restoration of most of the oxidized sulfhydryl groups and the disappearance of cytoskeletal aggregates, prevents the formation of surface blebs, and delays the occurrence of cell death.

Concerning microtubules, it is conceivable that the oxidation of thiol groups in tubulin and in microtubule-associated proteins (MAPs) could induce marked alterations of microtubule organization. Immunocytochemical studies performed in cultured mammalian cells have in fact demonstrated a complete depolymerization on microtubules following exposure to oxidative stress [6]. Previous studies carried out in several laboratories revealed that the oxidation of some of the 22 -SH residues in tubulin in addition to causing microtubule depolymerization also affect the biochemical process of microtubule formation from cytosolic precursors induced by GTP [20]. A complete inhibition of microtubule formation triggered by GTP has been observed when cytosolic precursors from oxidative stress-exposed cells were employed. Interestingly, this inhibition was completely reversed when DTT was included in the reaction medium, thus demonstrating a key role of thiol oxidation induced by oxidative stress [6].

Different Susceptibility of Nuclear Matrix Protein Sulfhydryl Groups to Oxidative Modifications

We described earlier the cytoskeleton as generally composed of three different classes of fibers or structures. Actually, this statement is not completely correct, since experimental evidence obtained in these past few years has indicated that, in addition to microfilaments, microtubules, and intermediate filaments, another class of structural elements exists in the cell, the nuclear matrix scaffold [10]. The nuclear matrix is bounded by an outer nuclear lamina connected to the cytoskeletal framework and is composed of a meshwork of inner filaments. The filaments range in diameter from 3 to 22 nm and are organized in a anastomosing network in which the nucleoli are embedded. The nuclear lamina is apparently composed of intermediate-type filaments and lines the nucleoplasmic surface of the nuclear envelope [16]. A key physiological role has been attributed to the nuclear matrix since it appears involved in the control of fundamental processes such as the regulation of DNA superhelicity, DNA replication, transcription, and RNA processing and transport.

We employed a procedure adapted from that originally developed by Staubenfield and Depper [28] to isolate nuclear matrix proteins from isolated rat hepatocytes. This procedure involves a sort of subsequent "biochemical dissection" of the whole cytoskeletal fraction isolated from starting cells that ultimately leads to the differentiation between an intermediate-filament fraction and a nuclear matrix fraction, on the basis of a different solubility of the proteins in an urea-free medium [33]. As reported in Table 4, it clearly appears that the amount of sulfhydryl groups present in nuclear matrix proteins is sig-

Table 4. Depletion of sulfhydryl groups in different cytoskeletal structures isolated from rat hepatocytes exposed to 2-methyl-1,4-naphthoquinone (menadione)

Fractions	Protein sulfhydryl groups (nmol/mg protein)	
	Control cells	Menadione-treated cells
Total cellular proteins	98 ± 11	54 ± 9
Whole cytoskeleton	84 ± 10	39 ± 9
DNAase-treated cytoskeleton	86 ± 10	42 ± 8
KCl-extracted cytoskeleton	97 ± 11	50 ± 9
Intermediate filaments	81 ± 10	41 ± 7
Nuclear matrix proteins	114 ± 8	101 ± 10

Isolated rat hepatocytes (10^6 cells/ml) were incubated for 60 min with or without 200 μM menadione and then processed for the sequential extraction of the fractions indicated. Protein concentration and quantitation of sulfhydryl groups were then performed. For experimental details on the procedure employed, see Vairetti et al. [33]

nificantly higher than that in all the other protein fractions. More interestingly, the -SH residues of nuclear matrix proteins were extremely resistant to oxidative depletion induced by the oxygen free radical generating compound menadione.

It is unlikely that the different susceptibility of this subset of cellular protein thiols to oxidation could be exclusively linked to a sort of inaccessibility of these proteins to the oxygen reactive species. In fact, despite their extremely short life, oxygen radicals easily reach the nucleus, where they induce a wide variety of DNA breakages. It is more conceivable to hypothesize that a well-organized defense system exists within the nucleus to prevent protein thiol oxidation. It is well-known, for instance, that a group of glutathione transferases and glutathione peroxidases is specifically located in the nucleus [29]. Moreover, recent work from our laboratory using multiparametric digitized video-microscopic analysis of cells loaded with fluorescent glutathione indicators has provided experimental support for a high intranuclear concentration of glutathione, approximately three times more than that found in the cytoplasm [9]. This intranuclear glutathione form is more resistant than the cytoplasmic one to the depletion caused by alkylating and oxidizing compounds.

Taken together, these findings suggest that nuclear matrix proteins are less sensitive to oxidative modifications of their sulfhydryl groups because they are much better protected from the oxidant insult, since they appear to be compartmentalized in a region of the cell in which GSH and glutathione-dependent defense-directed enzymes are particularly abundant.

Conclusions

During oxidative insults, both cell structure and function are dramatically affected. A key event in oxidative cell injury is represented by changes in the glutathione and protein thiol status that may ultimately cause disruption of ion homeostasis, perturbation of signal transduction, and depletion of energy stores. The cytoskeleton appears to be a critical target in oxidative injury, and alterations in it may significantly impair cell dynamics and generate morphological changes. However, the different cytoskeletal structures and polypeptides do not appear equally susceptible to oxidative modifications. This difference is related to their structural compartmentation and to the existence of specific defense systems.

It can easily be argued that the loss of antioxidant defenses or an increased rate of oxidant production (either experimentally induced or naturally occurring, like in the aging process) may affect this highly ordered structural and metabolic compartmentation and may lead to the oxidation of protein -SH residues critical for cell survival.

References

1. Becker PS, Cohen CM, Lux SE (1986) The effect of mild diamide oxidation on the structure and function of human erythrocyte spectrin. J Biol Chem 261: 4620 – 4628
2. Bellomo G, DiMonte D, Jewell SA, Richelmi, P, Orrenius S (1986) *t*-Butyl hydroperoxide toxicity in isolated hepatocytes: contribution of thiol oxidation and lipid peroxidation. J Biochem Toxicol 2: 13 – 21
3. Bellomo G, Mirabelli F, DiMonte D, Richelmi P, Thor H, Orrenius C, Orrenius S (1987) Formation and reduction of glutathione-protein mixed disulfides during oxidative stress. Biochem Pharmacol 36: 1313 – 1320
4. Bellomo G, Thor H, Orrenius S (1987) Alterations in inositol phosphate production during oxidative stress in isolated hepatocytes. J Biol Chem 262: 1530 – 1534
5. Bellomo G, Thor H, Mirabelli F, Richelmi P, Finardi G, Orrenius S (1989) Alterations in hepatocyte cytoskeleton during the metabolism of quinones. Adv Biosci 76: 21 – 35
6. Bellomo G, Mirabelli F, Vairetti M, Iosi F, Malorni W (1990) Cytoskeleton as a target in menadione-induced oxidative stress in cultured mammalian cells. Biochemical and immunocytochemical features. J Cell Physiol 143: 118 – 128
7. Bellomo G, Mirabelli F, Richelmi P, Malorni W, Iosi F, Orrenius S (1990) The cytoskeleton as a target in quinone toxicity. Free Rad Res Commun 8: 391 – 399
8. Bellomo G, Fulceri R, Albano E, Pompella A, Gamberucci A, Parola M, Benedetti A (1991) Ca^{2+}-dependent and independent mitochondrial damage in hepatocellular injury. Cell Calcium (in press)
9. Bellomo G, Vairetti M, Stivala L, Mirabelli F, Richelmi P, Orrenius S (1991) Demonstration of nuclear compartmentation of glutathione in hepatocytes. (in press)
10. Capco DG, Wan KM, Penman S (1982) The nuclear matrix: three dimensional architecture and protein composition. Cell 29: 847 – 858
11. Comporti M (1989) Three models of free radical-induced cell injuri. Chem Biol Interact 72: 1 – 56
12. Darnell J, Lodish H, Baltimore D (eds) (1990) Molecular cell biology. Scientific American Books, New York
13. DiMonte D, Ross D, Bellomo G, Eklow L, Orrenius S (1984) Alterations in intracellular thiol homeostasis during the metabolism of menadione by isolated hepatocytes. Arch Biochem Biophys 235: 334 – 342
14. Dolphin D, Poulson R, Avramovic O (eds) (1989) Glutathione: chemical, biochemical and medical aspects. Wiley, New York
15. Finch CE, Johnson TE (eds) (1990) Molecular biology of aging. Liss, New York
16. Gerace L (1986) Nuclear lamina and organization of nuclear architecture. Trends Biochem Sci 11: 443 – 446
17. Heller KB, Poser B, Haest CWM, Deutiche B (1984) Oxidative stress in human erythrocytes by iodate and periodate. Biochim Biophys Acta 777: 107 – 116
18. Ishikawa T, Sies H (1989) Glutathione as an antioxidant: toxicological aspects. In: Dolphin D, Poulson R and Avramovic O (eds) Glutathione: chemical, biochemical and medical aspects. Wiley, New York, pp 85 – 110
19. Jewell SA, Bellomo G, Thor H, Orrenius S, Smith MT (1982) Bleb formation in hepatocytes during drug metabolism is caused by disturbances in thiol and calcium ion homeostasis. Science 217: 1257 – 1259
20. Luduena RF, Roach MC, Jordan MA, Murphy DB (1985) Different reactivity of brain and erythrocyte tubulins toward a sulfhydryl group directed reagent that inhibits microtubule assembly. J Biol Chem 260: 1257 – 1264
21. Mirabelli F, Salis A, Marinoni V, Finardi G, Bellomo G, Thor H, Orrenius S (1988) Menadione-induced bleb formation in hepatocytes is associated with the oxidation of thiol groups in actin. Arch Biochem Biophys 264: 261 – 269
22. Mirabelli F, Salis A, Vairetti M, Bellomo G, Thor H, Orrenius S (1989) Cytoskeletal alterations in human platelets exposed to oxidative stress are modulated by oxidative and Ca^{2+}-dependent mechanism. Arch Biochem Biophys 270: 478 – 488

23. Nicotera P, Hartzell P, Baldi C, Svensson SA, Bellomo G, Orrenius S (1986) Cystamine induces toxicity in hepatocytes through the elevation of cytosolic Ca^{2+} and the stimulation of an non-lysosomal proteolytic system. J Biol Chem 261: 14628 – 14635
24. Nicotera P, Bellomo G, Orrenius S (1990) The role of Ca^{2+} in cell killing. Chem Res Toxicol 3: 484 – 494
25. Orrenius S, McConkey D, Bellomo G, Nicotera P (1989) Role of Ca^{2+} in toxic cell killing. Trends Pharmacol Sci 10: 281 – 285
26. Pompella A, Romani A, Benedetti A, Comporti M (1991) Loss of protein thiols and lipid peroxidation in allyl alcohol hepatotoxicity. Biochem Pharmacol (in press)
27. Reed DJ (1990) Glutathione: toxicological implications. Annu Rev Pharmacol Toxicol 30: 603 – 631
28. Staubenfield M, Depper W (1984) Preparation of nuclear matrices from cultured cells: fractionation of nuclei in situ J Cell Biol 98: 1886 – 1894
29. Thor H, Smith MT, Hartzell P, Bellomo G, Jewell SA, Orrenius S (1982) The metabolism of menadione (2-methyl-1,4-naphthoquinone) by isolated hepatocytes. J Biol chem 257: 12419 – 12425
30. Thor H, Mirabelli F, Salis A, Cohen GM, Bellomo G, Orrenius S (1988) Alterations in hepatocyte cytoskeleton caused by redox cycling and alkylating quinones Arch Biochem Biophys 266: 397 – 407
31. Tirmenstein MA, Nelson SD (1990) Acetaminophen-induced oxidation of protein thiols. J Biol Chem 265: 3059 – 3065
32. Ursini F, Bindoli A (1987) the role of selenium peroxidases in the protection against oxidative damage of membranes. Chem Phys Lipids 44 255 – 276
33. Vairetti M, Mirabelli F, Richelmi P, Bellomo G (1989) The biochemical dissection of the cytoskeleton reveals oxidative stress-induced alterations in intermediate filament-nuclear matrix scaffold of BT-20 cells. Med Biol Environ 17: 327 – 340

Unravelling the Molecular Defect in Alzheimer's Disease*

R. Prior, U. Mönning, C. L. Masters, and K. Beyreuther

Introduction

Dementia is of growing concern and social impact in the countries of the industrialized world where a strong rise in life expectancy has now led to increased interest in the biology of aging and in the pathology of age-associated diseases. Among these, Alzheimer's disease (AD) is the most frequent cause of dementia and as such already the fourth leading cause of death. Its pathological basis was defined by Alois Alzheimer in 1907 in a case report on a presenile demented patient in whom intracerebral deposits with histochemical properties typical for amyloid were the main histological finding [1]. These deposits are, as is characteristic for amyloid, congo red positive and green-red birefringent with polarized light and are found within neurons as neurofibrillary tangles, between neurons and glial cells as senile plaques, and within the walls of small cerebral and leptomeningeal blood vessels as vascular amyloid of congophilic angiopathy. The cerebral amyloid protein of AD is highly insoluble, and only in recent years have suitable purification techniques been developed. In 1984, Glenner and Wong [8] described the sequence of a major component of cerebrovascular amyloid, and in 1985 our laboratories analyzed the amyloid protein from senile plaques of AD and individuals with Down's syndrome (DS) [21]. In both instances, an almost identical polypeptide of 42/43 residues was found which is now termed βA4 protein according to its β-pleated sheet structure and its molecular weight of 4.5 kDa. Using synthetic βA4 and the respective antisera, βA4 has been confirmed as the amyloid protein of AD by its property of forming fibrils in vitro [11] and by its immunohistochemical detection in senile plaques and cerebrovascular amyloid [5, 34]. Intracerebral and cerebrovascular βA4 deposits are found not only in AD but also in DS [8, 21] and in the rare hereditary syndrome of cerebral hemorrhage with amyloidosis of the Dutch type (HCHWA-D). As in DS, in HCHWA-D the gradual development of dementia is a common clinical feature [10]; the neuropathological changes and the topographical distribution of βA4 are identical in AD and DS [18, 35].

*Financial support from the Deutsche Forschungsgemeinschaft (SFB317), the Bundesministerium für Forschung und Technologie, the Thyssen Foundation, the Fonds der Chemischen Industrie, and the National Health and Medical Research Council of Australia.

Reihe der Villa Vigoni, Band 1
Biology of Aging
Zwilling and Balduini (Eds.)

βA4 is derived from a larger glycosylated precursor protein (APP) that resembles a cell surface receptor and is controlled by the PAD gene (precursor of amyloid in AD and DS), also termed the APP gene [2], located on chromosome 21 [36]. Various alternatively spliced mRNAs have been identified, the main forms coding for APP molecules with 695, 751, and 770 amino acid residues which all contain the complete and partially membrane-inserted βA4 sequence [14, 15, 30, 45]. APP expression in transfected cell lines demonstrates the occurrence of secretory forms of APP695, 751, and 770 that are soluble. Secretory APP forms have also been detected in cerebrospinal fluid (CSF) [28, 49] and blood [29, 35].

Current approaches in the field of basic AD research center on four major topics: (a) the physiological role of APP, (b) the molecular mechanisms that lead to intracerebral βA4 deposits, (c) the recognition of AD-specific biochemical alterations in serum or CSF, and (d) the regulation of the PAD gene and of the alternative splicing of APP hnRNA. In the following, we will present a selection of current issues regarding points a – c, whereas the regulation, expression, and alternative splicing of the gene for APP is covered in the chapter by G. König et al., this volume.

Physiological Role of APP

APP is an evolutionarily highly conserved protein that shows strong homology among the known sequences of man, rat, mouse, and even *Drosophila* [25]. Its different mRNAs are moreover found in a wide variety of tissues derived from all three germ layers: APP695 is found predominantly in the brain, whereas APP751 and APP770, which both contain a 56-amino-acid insert with strong homology to the Kunitz type II family of serine protease inhibitors, are the most common isoforms in peripheral organs [9, 13]. The secretory forms of APP751, and probably also of APP770, are identical to protease nexin II originally isolated from fibroblasts by its property of forming complexes with epidermal growth factor binding protein [16, 27, 48]. The insert-containing forms of APP function in vitro as protease inhibitors, and APP751 has been localized to the α-granules of platelets, where it is released after thrombin stimulation [4, 47] and acts as a strong inhibitor of blood coagulation Factor XIa [41]. We have recently shown that another type of blood cell, i.e., peripheral blood lymphocytes, also produce APP751 [23]. Interestingly, APP mRNA or protein is not detectable in resting lymphocytes but is transcribed at a maximum level 24 h after stimulation with T-cell mitogens such as phytohemagglutinin or concanavalin A. Secretory APP751/770 may participate in the regulation of proteolytic activation cascades of specific physiological functions, e.g., in blood coagulation or lymphocyte activation. In the brain, APP is found in all types of neurons and also in astrocytic and ependymal glial cells and the choroid plexus epithelium [42, 46]. It is furthermore produced by leptomeningeal and intracerebral blood vessels [44]. APP colocalizes with synaptic markers such as synaptophysin and can be detected by scanning light as well as electron microscopy near synaptic clefts of the central nervous system (CNS) [3]. In peripheral nerves, APP is produced in the neuronal cell body and then transported by fast anterograde axonal transport to peripheral synaptic sites [17]. These observations indicate a role for APP695 in the maintenance of neuronal plasticity and the regulation of cell-cell interactions. Also, astrocytes and in particular reactive astrocytes produce large amounts of APP [39]. In order to investigate in situ the role of APP in proliferating brain cells of different origins, we studied the expression of APP in human brain tumors of various histological types by immunohistochemistry. A large fraction of glial tumors, e.g., astrocytomas and glioblastomas, were strongly immunoreactive for APP. When the APP content of these tumors is semiquantitatively compared with the grade of malignancy, a correlation is found (Fig. 1). Most semimalignant and malignant (WHO grades of malignancy III and IV), i.e., rapidly proliferating and highly invasive tumors, show expression of APP in the majority of their cells. In contrast, benign and semibenign (WHO grades I and II) tumors are far less positive for APP.

The expression of APP in highly invasive tumors is consistent with a role for APP in cell-cell interactions. Such a role is further supported by its cell-surface and membrane-spanning localization [6, 14, 38, 49] and by APP binding to extracellular matrix constituents [37].

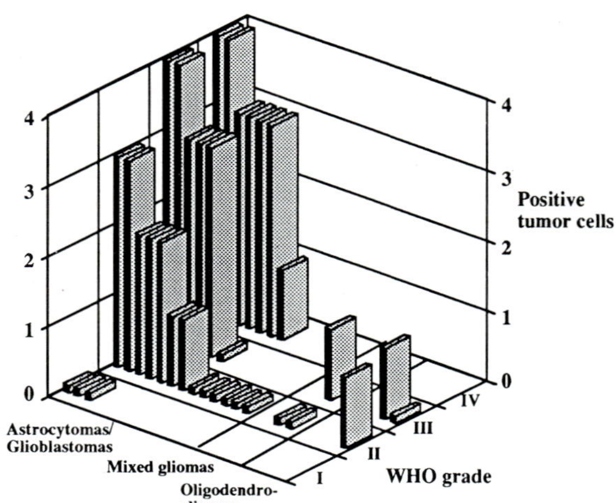

Fig. 1. Amyloid precursor protein (APP) expression in human gliomas. Immunoreactivity for APP was semiquantitatively graded on a rating scale with 0 = no positive cell, 1 = single positive cells, 2 = moderate fraction of positive cells (< 50 %), 3 = high fraction of positive cells (> 50 %), and 4 = very high fraction of positive cells (> 90 %). Malignant astrocytomas and glioblastomas (WHO grades III and IV) show strong expression of APP

Amyloidogenesis

Immunohistochemical methods have greatly improved the sensitivity of βA4 detection in brain tissue. An new morphological type of amyloid deposit has been identified, the "diffuse plaque" (Fig. 2), which consists of amorphous, extracellular βA4 that had not been recognized with the previously used congo red and silver impregnation techniques [34, 35, 50]. In AD, diffuse plaques are widely distributed throughout the grey matter of the CNS, including the basal ganglia, brain stem, cerebellum, and spinal cord [12, 26]; the latter locations were previously thought to be unaffected by AD pathology. Diffuse plaques appear at a time when neurofibrillary tangles, neuronal loss, microglial invasion, and reactive gliosis are still absent [34]. Their formation is an early event and at present the first histological marker during the evolution of AD [3]. As in other types of amyloidosis (for review, see [43]), the deposition of βA4 is highly tissue-specific. Whereas APP is almost ubiquitously found, relevant βA4 deposits occur only within the CNS. Moderate βA4 deposits are normally present in the aging brain [5] and probably constitute a presymptomatic pathological indication of AD [3]. It is therefore of primary interest, first, to characterize the molecular mechanisms that determine the CNS specifity of βA4 aggregation and, second, to identify the factors that induce the formation and finally lead to the accumulation of βA4 in AD. In addition to its localization in the brain parenchyma, βA4 is found in the outer and middle layers of small leptomeningeal blood vessels, which in contrast to the also amyloidotic brain capillaries are not in contact with cellular elements of the CNS. Leptomeningeal blood vessels are, however, surrounded by CSF.

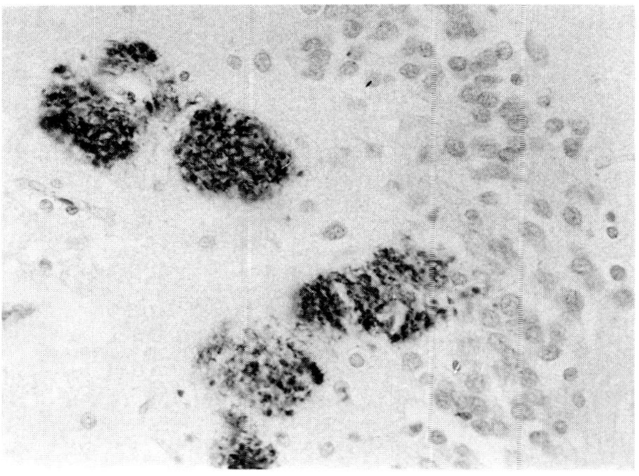

Fig. 2. Multiple diffuse βA4 amyloid deposits are visualized close to neurons of the hippocampal fascia dentata by immunostaining with βA4 amyloid polyclonal antiserum (counterstain with hematoxylin, ×150)

It is tempting to hypothesize the presence of a soluble "βA4 aggregation factor" or soluble forms of βA4 molecules or intermediate fragments of APP [20] including the entire βA4 sequence [6] specifically produced in the CNS and released into the CSF where they could infiltrate the walls of leptomeningeal blood vessels. Before βA4 amyloid can form, βA4 has to be proteolytically cleaved from APP. The proteolytic cleavage generating secretory APP occurs, however, within the βA4 sequence, producing nonamyloidogenic APP, and is therefore not involved in the pathological breakdown of APP to βA4 protein [7, 40]. A different, brain-specific pathological pathway with cleavage of the APP N-terminal and C-terminal to the βA4 sequence could explain the raised tendency of βA4 to intracerebral aggregation. Since isolated βA4 is highly insoluble and, at physiological pH and salt concentrations, readily aggregates [11], only minimal amounts of alternatively cleaved APP would lead to insoluble βA4 molecules. Furthermore, aggregated βA4 might be particular resistant in the brain, where the otherwise very efficient phagocytosis and proteolytic degradation by macrophages and cells of the reticuloendothelial system are replaced through microglial elements. Finally, a high ratio of APP695 mRNA to the protease inhibitor-containing APP751/770 mRNAs is found exclusively in brain tissue. This may be relevant for the postulated brain-specific pathological proteolytic processing of the respective proteins. The development of βA4 amyloidosis in vivo is, however, an extremely slow process, with an estimated time span of 30 years for the formation of senile plaques [3, 25]. The experimental generation of βA4 has to make use of the overexpression of APP or amyloidogenic APP fragments [6] to accelerate the natural rate of βA4 amyloid formation. In addition, laborious techniques such as the use of multiple antibodies or immunoelectron microscopy are needed for a sensitive and specific demonstration of newly formed βA4 fibrils. A very recent report describes for the first time the intracellular formation of fibrillary βA4 in cells transfected with cDNA coding for the C-terminal 100 residues of APP [20]. The future search for factors that inhibit the formation of βA4 depends on suitable in vivo models such as APP-overexpressing transgenic mice which are currently being investigated in our laboratory. A very promising in vivo model is the simulation of trisomy 21 (DS) in mice [32, 33]. It is unknown, whether the overexpression of APP due to an additional copy of chromosome 21 is the only mechanism that leads to the early and regular development of AD in patients with DS [35]. Cerebral transplants of trisomy 16 mice, the murine correspondent to trisomy 21, are positive for a large panel of cellular and extracellular AD markers and might be very useful to study in vivo different stages of βA4 amyloid formation and the pathogenetic relevance of different genes located on the human chromosome 21, e.g., the gene for the free radical scavenger superoxide dismutase.

Biochemical Markers of Alzheimer's Disease

The clinical diagnosis of AD is actually obtained by evaluation of the patient's history and a combination of psychological, neurological, and radiological examinations following criteria established in 1984 by the NINCDS-ADRDA work group [22]. It is made largely by exclusion of differential diagnoses such as multiinfarct dementia or cerebral neoplasia. The definitive diagnosis still depends on the post-mortem examination of brain tissue, an unfortunate situation for therapeutic clinical studies. Also, there is no possibility of separating preclinical from symptomatic AD patients; many comparative clinical evaluations are probably biased due to the long preclinical phase of AD, which cannot be recognized with the available examinations. A serum or CSF investigation with high specificity for AD would also be helpful for the identification of AD risk groups, the identification of risk factors closely associated with AD, and the evaluation of environmental influences during the progression of AD. AD-associated changes of intracerebral production, processing, or secretion of APP might be reflected by alterations of the APP levels and isoform ratios in the CSF. We therefore developed an enzyme-linked immunosorbent assay (ELISA) that measures the concentration of total APP in the CSF and analyzed the CSF content of APP in 22 AD and 10 neurologically normal control patients. In addition, we measured the quantities of APP695 and APP751/770 in the CSF by densitometric analysis of Western blots [31]. Both isoforms showed reduced CSF levels in the group of AD patients (Fig. 3). Lowered secretion of APP into the CSF might lead to long-term intracerebral accumulation of APP membrane proteins favoring βA4 amyloid deposition. Using our quantitative assays for APP, we hope now to determine the time course of APP secretion during different stages of AD and establish the pathogenetic and diagnostic relevance of such measurements.

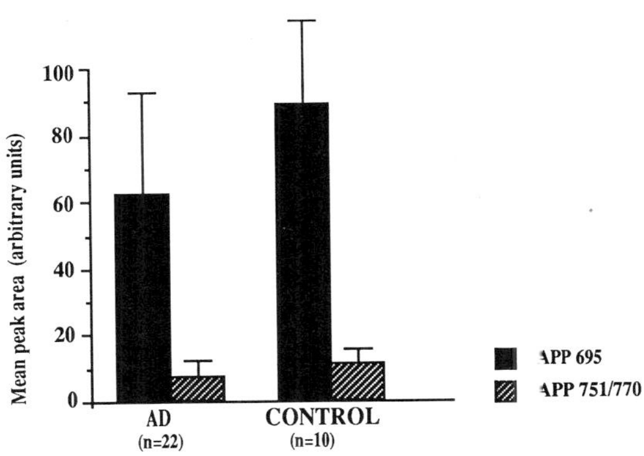

Fig. 3. Secretory amyloid precursor proteins, APP695 and APP751/770, measured by densitometric analysis of CSF Western blot autoradiographs. The mean APP content is expressed in arbitrary units, the mean total APP content (APP695 + APP751/770) of the control cases being set at 100. Total APP decreased from 100 ± 29.7 in the normal controls to 69.2 ± 35.7 in the Alzheimer's disease (*AD*) group ($P<0.025$). APP751/770 constituted a mean of 11 % in both groups

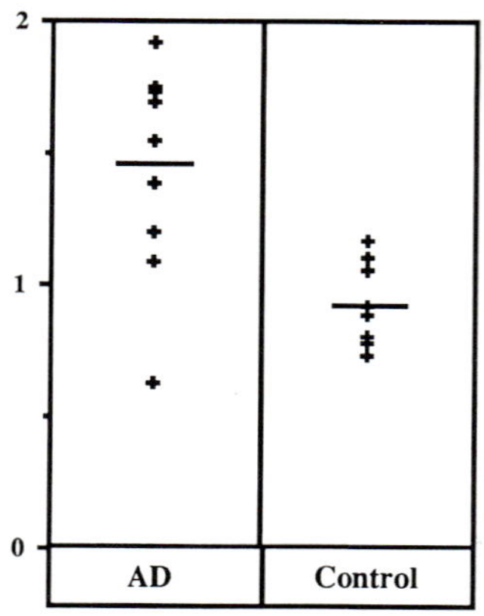

Fig. 4. βA4 autoantibodies in CSF of Alzheimer's disease (*AD*) patients and neurologically normal controls. The titer of the CSF autoantibodies was determined by an enzyme-linked immunoadsorbent assay and is expressed in arbitrary units as a ratio between the optical absorbance at 492 nm and the total IgG content of the respective CSF samples. Elevated immunoreactivity for βA4 is seen in the CSF of AD patients

Normal human blood sera and CSF contain antibodies that are reactive with synthetic βA4 protein. In a first analysis of 9 AD patients and 8 controls, we have shown an increase of these autoantibodies in the CSF of AD patients (Fig. 4), whereas the antibody titers in the serum were unchanged [24]. We are currently investigating the specificity of elevated levels of βA4-specific autoantibodies, particularly with regard to a distinction between normal and pathological titers. If βA4 antibodies should prove to be a valuable diagnostic tool for AD, a second and still more important step will be necessary: The identification of biochemical abnormalities which are primarily related to the abnormal processing of APP and which may allow us to recognize the pathological processing of APP long before the irreversible development of cerebral βA4 amyloidosis and dementia.

References

1. Alzheimer A (1907) Über eine eigenartige Erkrankung der Hirnrinde. All Z Psychiatr 64: 497 – 506
2. Beyreuther K, Masters CL (1990) Nomenclature of amyloid A4 proteins and their precursors in Alzheimer's disease and Down's syndrome. Neurobiol Aging 11: 66 – 68
3. Beyreuther K, Dyrks T, Multhaup G, Salbaum JM, Schubert W, Weidemann A, Masters CL (1990) Molecular genetics of dementia of Alzheimer's type: towards an early warning and treatment for individuals at risk. In: Fowler CJ, Carlson LA, Gottfries CG, Winblad B (eds) Biological markers in dementia of Alzheimer type. Smith-Gordon, London, pp 49 – 60
4. Bush AI, Martins RN, Rumble B, Moir R, Fullser S, Milward E, Currie J, Ames D, Weidemann A, Fischer P, Multhaup G, Beyreuther K, Masters CL (1990) The amyloid precursor protein of Alzheimer's disease is released by human platelets. J Biol Chem 265: 15977 – 15983

5. Davies L, Wolska B, Hilbich C, Multhaup G, Martins R, Simms G, Beyreuther K, Masters CL (1988) A4 amyloid protein deposition and the diagnosis of Alzheimer's disease: prevalence in aged brains determined by immunocytochemistry compared with conventional neuropathologic techniques. Neurology 38: 1688 – 1693
6. Dyrks T, Weidemann A, Multhaup G, Salbaum JM, Lemaire H-G (1988) Identification, transmembrane orientation and biogenesis of the amyloid A4 precursor of Alzheimer's disease. EMBO J 7: 949 – 957
7. Esch GS, Keim PS, Beattie EC, Blacher RW, Culwell AR, Oltersdorf T, McClure D, Ward PJ (1990) Cleavage of amyloid β peptide during constitutive processing of its precursor. Science 248: 1122 – 1124
8. Glenner GG, Wong CW (1984) Initial report of the purification and characterization of a novel cerebrovascular amyloid protein. Biochem Biophys Res Commun 120: 885 – 890
9. Golde TE, Estus S, Usiak M, Younkin LH, Younkin SG (1990) Expression of β amyloid protein precursor mRNAs: recognition of a novel alternatively spliced form and quantitation in Alzheimer's disease using PCR. Neuron 4: 253 – 267
10. Haan J, Lanser JBK, Zijderveld I, Vanderdoes IGF, Roos RAC (1990) Dementia in hereditary cerebral hemorrhage with amyloidosis Dutch type. Arch Neurol 47: 965 – 967
11. Hilbich C, Kisters-Woike B, Reed J, Masters CL, Beyreuther K (1991) Aggregation and secondary structure of synthetic amyloid βA4 peptides of Alzheimer's disease. J Mol Biol 218: 149 – 163
12. Joachim CL, Morris JH, Selkoe DJ (1989) Diffuse senile plaques occur commonly in the cerebellum in Alzheimer's disease. Am J Pathol 135: 309 – 319
13. Kang J, Müller-Hill B (1990) Differential splicing of Alzheimer's disease amyloid A4 precursor RNA in rat tissues: preA4695 is predominantly produced in rat and human brain. Biochem Biophys Res Commun 166: 1192 – 1200
14. Kang J, Lemaire HG, Unterbeck A, Salbaum JM, Masters CL, Grzeschik KH, Multhaup G, Beyreuther K, Müller-Hill B (1987) The precursor of Alzheimer's disease amyloid A4 protein resembles a cell-surface receptor. Nature 325: 733 – 736
15. Kitaguchi N, Takahashi Y, Tokushima Y, Shiojiri S, Ito H (1988) Novel precursor of Alzheimer's disease amyloid protein shows protease inhibitory activity. Nature 331: 530 – 532
16. Knauer DJ, Cunningham DD (1982) Epidermal growth factor carrier protein binds to cells via a complex with released carrier protein nexin. Proc Natl Acad Sci USA 79: 2310 – 2314
17. Koo EH, Sisodia SS, Archer DR, Martin LJ, Weidemann A, Beyreuther K, Fischer P, Masters CL, Price DL (1990) Precursor of amyloid protein in Alzheimer disease undergoes fast anterograde axonal transport. Proc Natl Acad Sci USA 87: 1561 – 1565
18. Mann DMA, Jones D, Prinja D, Purkiss MS (1990) The prevalence of amyloid (A4) protein deposits within the cerebral cortex in Down's syndrome and Alzheimer's disease. Acta Neuropathol 80: 318 – 327
19. Marayuma K, Terakado K, Usami M, Yoshikawa K (1990) Formation of amyloid-like fibrils in COS cells overexpressing part of the Alzheimer amyloid protein precursor. Nature 347: 566 – 569
20. Masters CL and Beyreuther K (1989) Amyloid A4 protein and its precursor in Down's syndrome and Alzheimer's disease. N Engl J Med 321: 1197
21. Masters CL, Simms G, Weinman NA, Multhaup G, McDonald BL, Beyreuther K (1985) Amyloid plaque core protein in Alzheimer's disease and Down's syndrome. Proc Natl Acad Sci USA 82: 4245 – 4249
22. McKhann G, Drachmann D, Folstein M, Katzman R, Price D, Stadlan EM (1984) Clinical diagnosis of Alzheimer's disease: report of the NINCDS-ADRDA work group under the auspices of department of health and human services task force on Alzheimer's disease. Neurology 34: 939 – 944
23. Mönning U, König G, Prior R, Mechler H, Schreiter-Gasser U, Masters CL, Beyreuther K (1990) Synthesis and secretion of Alzheimer amyloid βA4 precursor protein by stimulated human peripheral blood lymphocytes. FEBS Lett 277: 261 – 266

24. Mönning U, Schreiter-Gasser U, Hilbich C, Bunke D, Prior R, Masters CL, Beyreuther K (1991) Alzheimer amyloid βA4 protein-reactive antibodies in human sera and CSF. In: Iqbal K, McLachlan DRC, Winblad B, Wisniewski HM (eds) Alzheimer's disease: basic mechanisms, diagnosis and therapeutic strategies. Wiley, Chichester, pp 557 – 563
25. Müller-Hill B, Beyreuther K (1989) Molecular biology of Alzheimer's disease. Annu Rev Biochem 58: 287 – 307
26. Ogomori K, Kitamoto T, Tateishi J, Sato Y, Suetsugu M, Abe M (1989) β-Protein is widely distributed in the central nervous system of patients with Alzheimer's disease. Am J Pathol 134: 243 – 251
27. Oltersdorf T. Fritz LC, Schenk DB, Lieberburg I, Johnson-Wood KL, Beattie EC, Ward PJ, Blacher RW, Dovey HF, Sinha S (1989) The secreted form of the Alzheimer's amyloid precursor protein with the Kunitz domain is protease nexin-II. Nature 341: 144 – 147
28. Palmert MR, Podlisny MB, Witker DS, Oltersdorf T, Younkin LH, Selkoe DJ, Younkin SG (1989) The β-amyloid protein precursor of Alzheimer disease has soluble derivatives found in human brain and cerebrospinal fluid. Proc Natl Acad Sci USA 86: 6338 – 6342
29. Podlisny MB, Mammen AL, Schlossmacher MG, Palmert MR, Younkin SG, Selkoe DJ (1990) Detection of soluble forms of the β-amyloid precursor protein in human plasma. Biochem Biophys Res Commun 167: 1094 – 1101
30. Ponte P, Gonzalez-DeWhitt P, Schilling J, Miller J, Hsu D, Greenberg B, Davis K, Wallace W, Lieberburg I, Fuller F, Cordell B (1988) A new A4 amyloid mRNA contains a domain homologous to serine proteinase inhibitors. Nature 311: 525 – 527
31. Prior R, Mönning U, Schreiter-Gasser U, Weidemann A, Blennow K, Gottfries CG, Masters CL, Beyreuther K (1991) Quantitative changes in the amyloid βA4 precursor protein in Alzheimer cerebrospinal fluid. Neurosci Lett 124: 69 – 73
32. Richards S-J, Waters JJ, Rogers DC, Martel FL, Sparkman DR, White CL, Beyreuther K, Masters CL, Dunnett SB (1990) Hippocampal grafts derived from embryonic trisomy 16 mice exhibit amyloid (A4) and neurofibrillary pathology. In: Dunnet SB, Richards S-J (eds) Neural transplantation. From molecular basis to clinical applications. Elsevier, Amsterdam, pp 215 – 223 (Progress in brain research, vol 82)
33. Richards S-J, Waters JJ, Wischik CM, Abraham CR, Sparkman DR, White CL, Beyreuther K, Masters CL, Dunnett SB (1991) Transplants of mouse trisomy 16 hippocampus provide an in vivo model of the neuropathology of Alzheimer's disease. EMBO J 10: 297 – 303
34. Rozemuller JM, Eikelenboom P, Stam FC, Beyreuther K, Masters L (1989) A4 protein in Alzheimer's disease: primary and secondary cellular events in extracellular amyloid deposition. J Neuropathol Exp Neurol 48: 674 – 691
35. Rumble B, Retallack R, Hilbich C, Simms G, Multhaup G, Martins R, Hockey A, Montgomery P, Beyreuther K, Masters CL (1989) Amyloid A4 protein and its precursor in Down's syndrome and Alzheimer's disease. N Engl J Med 320: 1446 – 1452
36. Salbaum JM, Weidemann A, Lemaire HG, Masters CL, Beyreuther K (1988) The promotor of Alzheimer's disease amyloid A4 precursor gene. EMBO J 7: 2807 – 2813
37. Schubert D, LaCorbiere M, Saitoh T, Cole G (1989) Characterization of an amyloid β precursor protein that binds heparin and contains tyrosine sulfate. Proc Natl Acad Sci USA 86: 2066 – 2069
38. Shivers B, Hilbich C, Multhaup G, Salbaum JM, Beyreuther K, Seeburg PH (1988) Alzheimer's disease amyloidogenic glycoprotein: expression pattern in rat brain suggests a role in cell contact. EMBO J 7: 1365 – 1370
39. Siman R, Card JP, Nelson RB, Davis LG (1989) Expression of β-amyloid precursor protein in reactive astrocytes following neuronal damage. Neuron 3: 275 – 285
40. Sisodia SS, Koo EH, Beyreuther K, Unterbeck A, Price DL (1990) Evidence that β-amyloid protein in Alzheimer's disease is not derived by normal processing. Science 248: 492 – 495
41. Smith RP, Higuchi DA, Broze GJ Jr (1990) Platelet coagulation XIa-inhibitor, a form of Alzheimer amyloid precursor protein. Science 248: 1126 – 1128

42. Stern RA, Otvos L Jr, Trojanowski JQ, Lee VM (1989) Monoclonal antibodies to a synthetic peptide homologous with the first 28 amino acids of Alzheimer's disease beta-protein recognize amyloid and diverse glial and neuronal cell types in the central nervous system. Am J Pathol 134: 973 – 978
43. Stone MJ (1990) Amyloidosis: a final common pathway for protein deposition in tissues. Blood 75: 531 – 545
44. Tagliavini F, Ghiso J, Timmers WF, Giaccone G, Bugiani O, Frangione B (1990) Coexistence of Alzheimer's amyloid precursor protein and amyloid protein in cerebral vessel walls. Lab Invest 62: 761
45. Tanzi RE, McClatchey AI, Lamperti ED, Villa-Komaroff L, Gusella JF, Neve RL (1988) Protease inhibitor domain encoded by an amyloid protein precursor mRNA associated with Alzheimer's disease. Nature 311: 528 – 530
46. Tu GF, Cole T, Southwell BR, Schreiber G (1990) Expression of the genes for transthyretin, cystatin C and βA4 amyloid precursor protein in sheep choroid plexus during development. Dev Brain Res 55: 203 – 208
47. Van Nostrand WE, Schmaier AH, Farrow JS, Cunningham DD (1990a) Protease nexin-II (amyloid β-protein precursor): a platelet alpha-granule protein. Science 248: 745 – 748
48. Van Nostrand WE, Wagner SL, Suzuki M, Choi BH, Farrow JS, Geddes JW, Cotman CW, Cunningham DD (1990b) Protease nexin-II, a potent antichymotrypsin, shows identity to amyloid beta-protein precursor. Nature 341: 546 – 549
49. Weidemann A, König G, Bunke D, Fischer P, Salbaum JM, Masters CL, Beyreuther K (1989) Identification, biogenesis and localization of precursors of Alzheimer's disease A4 amyloid protein. Cell 57: 115 – 126
50. Yamaguchi H, Hirai S, Morimatsu M, Shoji M, Ihara Y (1988) A variety of cerebral amyloid deposits in the brain of the Alzheimer-type dementia demonstrated by β protein immunostaining. Acta Neuropathol 76: 541 – 549

Expression of Alzheimer's Amyloid Gene in Development, Aging, and Alzheimer's Disease*

G. König, C. L. Masters, and K. Beyreuther

Introduction

Alzheimer's disease (AD) is an irreversible neurodegenerative disorder of the human central nervous system [1, 32] caused by neuronal dysfunction. It is correlated with massive deposition of fibrillar aggregates of the amyloid βA4 protein. These amyloid deposits occur in the brain as intracellular neurofibrillary tangles, extracellular amyloid plaque cores, and cerebrovascular amyloid [7, 12, 20, 21, 23, 42 – 44, 46, 47]. The βA4 peptide consists of 42 – 43 residues and is synthesized as part of a larger precursor protein (APP, amyloid precursor protein) [16, 25, 31, 56, 63, 69].

The gene coding for APP maps to the long arm of chromosome 21 [25, 31, 56, 63]. So far, five different transcripts have been identified, which arise from the single APP gene through alternative splicing. Four transcripts with sizes of 3.4 – 3.6 kb code for precursor proteins with 695, 714, 751, and 770 amino acids (APP695, APP714, APP751 and APP770 [30, 33, 55, 64] (Fig. 1). The APP751 protein differs from APP695 by an additional domain of 56 amino acids which shows homology to the Kunitz type II serine protease inhibitors (KPI domain, Fig. 1) This domain was shown to function in vitro as an inhibitor of several proteases [33, 49]. APP770 contains an additional domain of 19 amino acids following the KPI domain, showing homology to the MRC-Ox-2 antigen [9, 47]. APP714 contains this short domain of 19 amino acids but not the KPI domain [24, 30]; and this transcript has so far only been described as a low abundant splice form, and its relevance remains to be determined. Both of these domains are encoded by two adjacent exons on the APP gene, namely exon 7 (168 bp) and exon 8 (57 bp) [33, 38]. The four largest precursor proteins are synthesized as N- and O-glycosylated integral transmembrane proteins (Fig. 1) [16, 67]. Proteolytic processing of these proteins in the Golgi apparatus leads to the secretion of the extracellular domains (Fig. 1) [67]. Recently a growth factor-like activity has been assigned to the secreted form of APP695 [58]. The secreted form of APP751 and possibly of APP770, both of which contain the KPI domain, are presumably identical

*This work was supported by grants from the Deutsche Forschungsgemeinschaft through SFB 317, the Bundesministerium für Forschung und Technologie, the Thyssen Foundation, and the Fonds der Chemischen Industrie. Gerhard König is a fellow of the Boehringer Ingelheim Fonds.

Reihe der Villa Vigoni, Band 1
Biology of Aging
Zwilling and Balduini (Eds.)

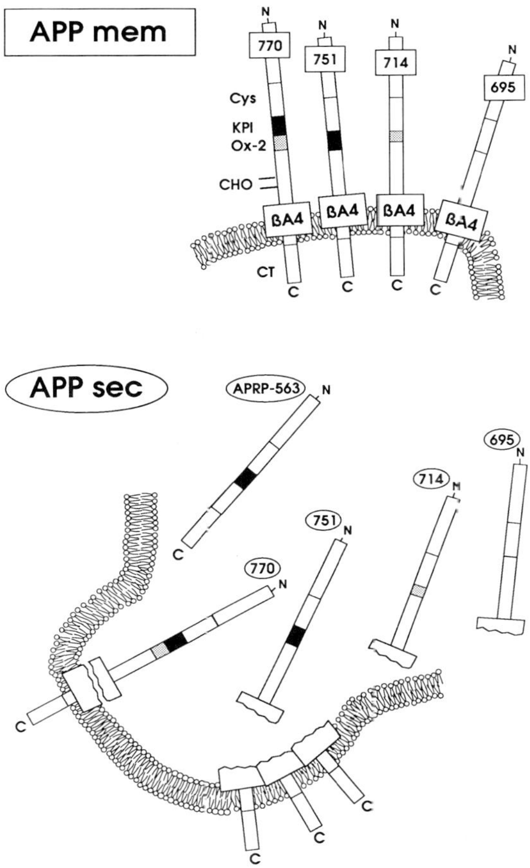

Fig. 1. Amyloid precursor proteins (APPs). *Upper panel* shows the membrane-anchored APP variants and their respective membrane topology. The Cys-rich domain (*Cys*), the Kunitz-type protease inhibitor (*KPI*), the domain with homology to the MRC-Ox-2 antigen (*Ox-2*), the carbohydrate domain (*CHO*), and the cytoplasmic part (*CT*) are indicated. *Lower panel* shows the secreted parts (*sec*) of the APP variants and the APRP563 variant. Secretion of APP leads to cleavage of the βA4 peptide, which therefore cannot be the pathological process leading to the accumulation of βA4 peptides

with protease nexin II [34, 49]. The Ox-2 domain of APP770/714 might play a role in the interaction of cells with the immune system.

A shorter transcript of 2.0 kb consisting of only the first 13 exons, including exon 7, and an additional Alu-repeat-like sequence on the 3' end was detected in the cerebral cortex of a patient with AD as well as in control brains [13]. The resulting protein APRP563 (amyloid precursor related protein) lacks the transmembrane domain and the βA4 amyloid region encoded by exons 16 and 17. This splice form of the APP gene is nonamyloidogenic, i.e., it does not contribute to the process of βA4 amyloid deposition since it lacks the βA4 portion (see Fig. 1).

The promoter of the gene coding for APP exhibits the typical features of a so-called housekeeping gene [4, 59]. It has a high GC content, multiple transcription initiation sites, and no TATA box. In addition to its constitutive expression it has several putative regulatory elements which might be responsible for a transcriptional regulation of the APP gene. Multiple acute phase elements and a heat-shock element could be involved in expression of the APP gene in response to inflammatory as well as stress conditions [66]. The role of potential AP-1 and SP-1 transcription factor binding sites of the APP promoter has yet to be determined. The expression of the APP gene is ubiquitous, which is consistent with its definition as a house keeping gene, being strongest in the brain and kidney [64]. In situ hybridizations as well as Northern blot analysis of brain and peripheral tissues demonstrated a raised expression of the two longer APP transcripts (APP751/770) in peripheral organs, whereas the short APP695 mRNA is the dominant form in neuronal tissues [48, 64]. A developmental regulation of the splicing pattern was indicated by results which showed that the APP695 mRNA is more strongly expressed in fetal brain than in adult human brain (see below; [64]). A number of recent reports address the question of whether one or more of the APP transcripts is up- or down-regulated in respect to AD; however, up to now the available data are still contradictory [29, 36, 37, 48, 51, 62]. In order to establish the expression pattern of the APP gene transcripts in developing human brain and in adult brain, we studied the splicing pattern of the different splice forms in fetal as well as in adult brains of individuals without neurological diseases and of different ages. The data obtained from the adult control brains were then compared with those from the cortex of individuals with AD. The outcome of this comparison led us to investigate further the expression of the APP gene during differentiation.

We used the human neuroblastoma cell line subclone SY5Y, which can be induced to differentiate by nerve growth factor (NGF), retinoic acid (RA), forskolin, and other factors [11, 52, 57]. Treatment of SY5Y cells with these agents for several days leads to neurite extension, growth arrest, and aggregation of the cells to pseudoganglia. Those differentiated neurons allowed us to study the process of differentiation in a simple model system. We analyzed the effects of inducers of differentiation on the expression and splicing pattern of the APP gene. In addition to Northern blot analysis, we applied a sensitive S1 nuclease protection assay, which enabled us to assess and quantify the major amyloidogenic APP transcripts, APP695, APP751, and APP770, in a single experiment [5, 65].

Here we report that a change of the splicing pattern in favor of APP751 and APP770 mRNA is observed in the control human adult cortex, compared with a developing fetal brain. The results obtained from patients with AD do not indicate a significant difference in the splicing ratio when compared with the aged-matched controls. In cell culture, differentiation is associated with an increase in APP gene expression. The response of the APP gene to NGF and RA achieves its maximum extent several days after administration and is correlated with the process of differentiation. Differentiation of neuroblastoma cells with RA resulted in the greatest increase of APP gene expression. We also observed a pronounced change of the splicing pattern of the APP transcripts in cells treated with RA, a pattern that is shifted towards the one observed in a developing fetal brain. In contrast, untreated neuroblastoma cells exhibit a splicing pattern that resembles that of fully differentiated neurons in the cortex of adult brain.

Results

S1 Nuclease Protection Assay

Previously published studies on the expression of the APP gene were based on Northern blot hybridizations or in situ hybridizations [10, 28, 40, 48, 51, 62, 64]. The sizes of the major APP mRNAs coding for the amyloidogenic membrane-bound precursor proteins vary from 3.4 to 3.6 kb, a range over which Northern blots give only limited size resolution [31]. To analyze quantitatively the different APP mRNAs, subsequent hybridization experiments of the same filter with different oligonucleotide probes are necessary. In order to achieve a more sensitive method to evaluate the amount of each of the APP transcripts, we developed a sensitive S1 nuclease protection assay that allows the detection and resolution of the three major amyloidogenic APP transcripts in a single experiment [5, 35]. To measure quantitatively the amount of total APP transcripts, a S1 nuclease protection assay as well as Northern blot hybridizations were performed.

A 2.7-kb cDNA clone containing part of the APP770 mRNA was isolated from a lambda gt10 human fetal brain library. A 826-bp cDNA fragment spanning the KPI and Ox-2 domain, from position +320 at the 5'end of our cDNA clone where an adaptor (*Eco*RI-*Xho*I-*Sac*I-*Sal*I) was inserted to position +1135, an *Xho*I restriction site, was subcloned into the phage M13mp18 which resulted in the recombinant phage M8.11 [31, 33, 35, 68] (Fig. 2a,b). This cDNA fragment spans the coding region for part of the cysteine domain, the acidic domain, the protease inhibitor domain, as well as the domain which shows homology to the MRC-Ox-2 antigen [47] (Fig. 2a).

The recombinant phage M15.36 was constructed through subcloning of a 682-bp *Xho*I/*Sac*I cDNA fragment from position +1135 to position +1817 of the APP770 cDNA into the phage M13mp19 (Fig. 2a,c). This cDNA fragment comprises the major part of the coding region for the carbohydrate domain [32]. Single-stranded DNA was prepared from the recombinant M13 clones M8.11 and M15.36 and used to synthesize uniformly radioactively labelled, single-stranded DNA probes which are complementary to the mRNA of the APP cDNAs. Probe M8.11 was subsequently used for the S1 nuclease protection experiments to quantify the ratio of APP transcripts. Excess probe was hybridized to total RNA, and the resulting hybrids were digested with S1

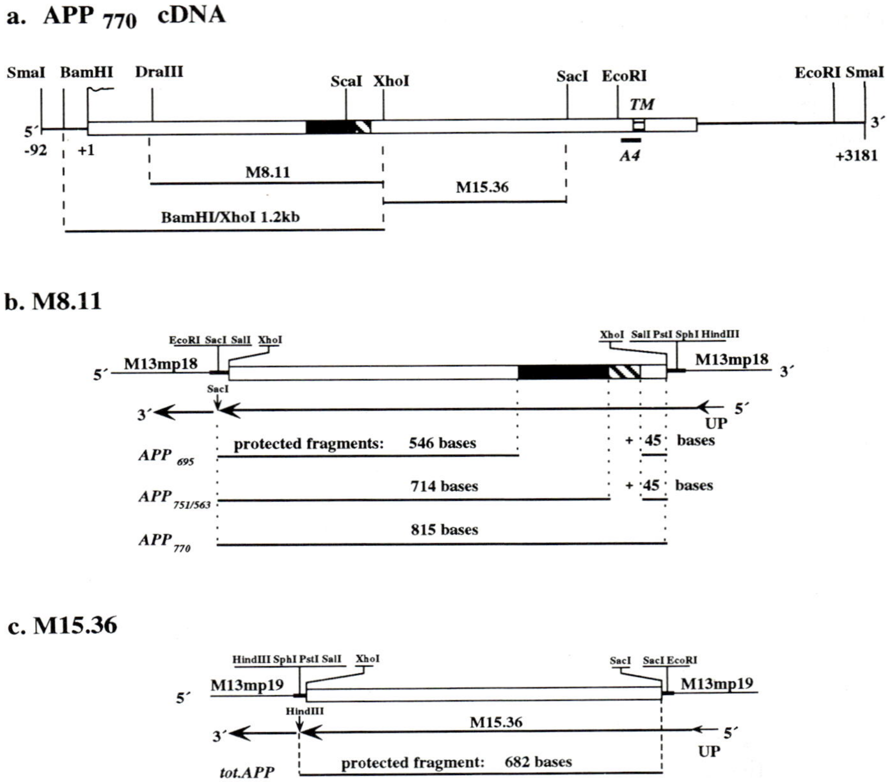

Fig. 2. **a** Amyloid precursor protein (APP) cDNA and the orientation of the partial cDNA probes described in the text. The exon coding for the Kunitz-type protease inhibitor (KPI) domain is denoted by the *black box*, the exon coding for the domain with homology to Ox-2 is denoted by the *hatched box*. *A4* stands for the region on the cDNA coding for the βA4 peptide, *TM* for the region coding for the transmembrane part of APP. Relevant restriction sites on the cDNA are indicated. **b** Recombinant M13 phage M8.11. The S1 probe M8.11 is generated from the universal primer (*UP*) and subsequently sized to 888 bases with restriction enzyme *Sac*I. The resulting protected fragments after hybridization and S1 nuclease digestion for the major amyloidogenic APP mRNAs are indicated. **c** Recombinant M13 phage M15.36. The S1 probe M15.36 is generated from the universal primer (*UP*) and subsequently sized to 682 bases with restriction enzyme *Hin*dIII. The resulting protected fragment after hybridization and S1 nuclease digestion for the major amyloidogenic APP mRNAs is indicated

nuclease [5]. Hybrids of APP770 mRNA with probe M8.11 give a fully protected, labelled fragment of 815 bases. Hybrids of APP751 mRNA with probe M8.11 have a single-stranded DNA loop corresponding to the exon coding for the domain with homology to the MRC-Ox-2 antigen. Digestion of these hybrids with S 1 nuclease results in protected fragments of 714 and 45 bases. Hybrids of APP695 mRNA with probe M8.11 have a larger single-stranded DNA loop, which corresponds to the two exons of the protease inhibitor domain and the domain with homology to the MRC-Ox-2 antigen. Digestion of these hybrids with S1 nuclease leaves protected fragments of 546 and 45 bases (Fig. 2b). The shorter, nonamyloidogenic APRP563 mRNA also gives rise to protected fragments of 714 bases, as is the case for APP751 mRNA, and was distinguished from the latter by a S1 nuclease protection assay using S1 probe M15.36.

Probe M15.36 was used to quantify the total amount of the three major amyloidogenic APP transcripts in our S1 assays. Hybrids of the amyloidogenic APP695, APP751, and APP770 mRNA all include a fully protected, labelled fragment of 682 bases (Fig. 2c). Hybridization with the APRP563 transcript resulted in a protected fragment of 551 bases [13]. We were, however, unable to detect this secreted APP mRNA form in SY5Y cells given the sensitivity of our assay. The protected fragments of probes M8.11 and M15.36 were resolved through denaturing polyacrylamide gel electrophoresis. The indicative radioactive fragments were detected by autoradiography and quantitated by densitometry, using exposure times dependent on the strength of the signals.

Splicing Pattern in Human Cortex

Earlier studies indicated a strong expression of APP695 mRNA in fetal brain compared with adult brain tissue [64]. To establish the exact expression pattern of the APP gene in a developing brain, we analyzed the RNA from cortical brain tissue of a 20-week-old fetus. Total RNA was prepared and a S1 nuclease protection assay performed with probe M8.11 [8]. After resolving the protected fragments on a denaturing polyacrylamide gel and subsequent densitometric scanning of the autoradiographs, we observed the splicing pattern shown in Fig. 3. APP695 mRNA was by far the dominant transcript with more than 90 % of total APP mRNA. The relative amounts of APP751 and APP770 mRNA were less than 9 % and 1 %, respectively. The signal corresponding to APP770 mRNA was only visible after a strong overexposure of the autoradiograph (data not shown). The responsiveness of the X-ray film to this radioactive signal is therefore no longer in a linear range, leading to an underestimate of the fetal splice ratio. The actual amount of APP695 mRNA is thus even higher and APP770 mRNA, lower than the values given here.

In contrast to fetal brain, APP695 mRNA is reduced to values of about 48 % of total APP mRNA for the younger control cases and ranges between approximately 40 % and 20 % for the older control cases (Fig. 3). This represents a reduction by a factor of 3 – 4 in the older control cases compared with fetal brain. APP751 mRNA becomes the dominant transcript in all control individuals examined (Fig. 3), with values ranging from approximately 48 % to 49 % in the younger adult cases. This increase of APP751 mRNA becomes more apparent in the older control cases (69 – 92 years; Fig. 3). Taken together, a five- to eightfold increase in the relative amounts of APP751 mRNA

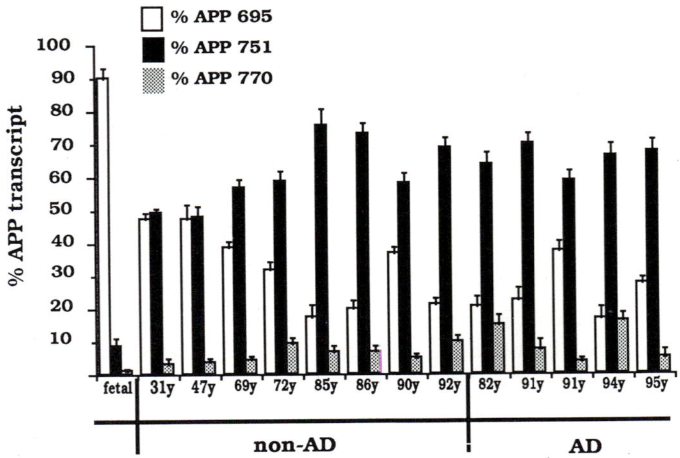

Fig. 3. Relative amyloidogenic amyloid precursor protein (APP) transcript amount in post-mortem tissues of human fetal brain, young adult Alzheimer's disease sufferer (*AD*), and age-matched, *non-AD* temporal cortex

is found in adult control brains during aging when compared with the splicing pattern in a developing fetal brain. The longest splice form, APP770 mRNA, represents from 3.5 % up to 10 % of total APP transcripts, which corresponds to an increase of about four- to tenfold when compared with the relative values of APP770 mRNA in fetal brain. These data are indicative of a developmentally regulated splicing of the APP hnRNA in the central nervous system and suggest that the corresponding proteins participate in developmental processes and aging.

The data obtained for the splicing pattern in the cortex of patients with AD varied more strongly than in the controls. Whereas APP695 mRNA is reduced to values of approximately 20 % – 40 % (Fig. 3), which is also found for our aged-matched control cases, APP770 mRNA is more abundant in two patients with AD. This increase of APP770 mRNA to about 16 % – 17 % of total APP mRNA is, however, not a consistent feature of the total of 8 patients examined with a pathologically confirmed diagnosis of AD (data not shown). As a first step towards an understanding of this documented change in the splicing pattern of the APP gene during development and subsequent aging, we turned to an in vitro cell culture system and analyzed the splicing pattern of undifferentiated and differentiated neuronal cells.

Expression of the Amyloid Precursor Protein Gene in SY5Y Cells Treated with Nerve Growth Factor

In order to obtain a homogenously growing population, the cells were kept in an early logarithmic phase at low density and passaged every 4th day. The neu-

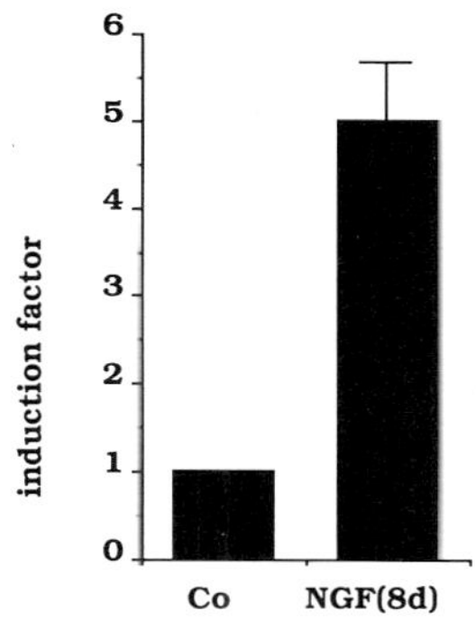

Fig. 4. Induction of total amyloid precursor protein gene expression in SY5Y neuroblastoma cells after differentiation with nerve growth factor (*NGF*)

roblastoma cell line SY5Y was treated with NGF (20 ng/ml) to induce differentiation [53]. After 6 – 8 days in culture in the presence of NGF, the majority of SY5Y cells extend neurites, cease multiplying, aggregate, and have electrically excitable membranes [53]. They resemble closely the phenotype of a primary culture of sympathetic ganglia after treatment with NGF. In the light of our previous findings which suggested an involvement of APP proteins in differentiation, we were then interested in the expression pattern of the APP gene in this cell line following its differentiation in vitro. SY5Y cells were cultured in the presence of NGF for 8 days, after which a morphological differentiation was observed. Total RNA was prepared, analyzed for integrity through denaturing agarose gel electrophoresis, and subsequently used for S1 nuclease analysis with probe M8.11 as well as with probe M15.36. In the differentiated cells after 8 days of NGF treatment, we found a fivefold higher expression of the APP gene with probe M15.36 [35] (Fig. 4). This is in agreement with the results reported for the in vivo situation of the APP gene in the developing hamster brain [46]. This response of the APP gene to the differentiating agent NGF is maximal after 8 days. The splice ratios determined with M8.11 for the SY5Y cells treated with NGF did not show a difference in the relative amounts of the APP transcripts at any of the time points chosen for analysis. In undifferentiated as well as differentiated SY5Y cells, the APP695 transcript accounts for about 49 %, the APP751 transcript for about 41 %, and the APP770 transcript for about 10 % of total APP mRNA (Fig. 5). Undifferentiated cells harvested at low density show a slightly higher ratio of APP695 mRNA of approximately 54 % (see below). We conclude that differentiation of the neuroblastoma cell line SY5Y with NGF is associated with a fivefold increase in APP gene expression but without change in the splicing pattern.

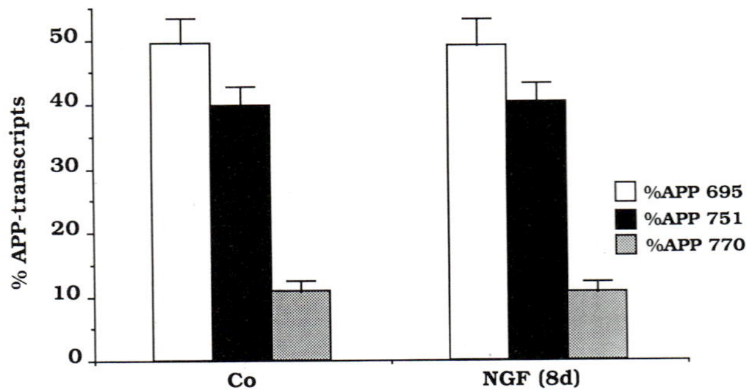

Fig. 5. Relative amyloidogenic amyloid precursor protein (*APP*) transcript amount in SY5Y neuroblastoma cells before and after differentiation with nerve growth factor (*NGF*)

Expression of the Amyloid Precursor Protein Gene in SY5Y Cells After Differentiation with Retinoic Acid

It has been reported that RA ist a potent inducer of differentiation in neuroblastoma cells [50]. We therefore tested this morphogen and investigated its effects on the expression of the APP gene. SY5Y cells were kept in early logarithmic growth phase and passaged every 4th day. When they were cultured in the presence of RA (10 μM), the majority developed the characteristic phenotype of differentiated neurons after 4–5 days. Between days 6 and 12 the typical morphology of differentiated neuroblastoma cells could be maintained. After the 2nd week, more and more flattened undifferentiated cells overgrew the differentiated ones, leading to a morphologically heterogenous population. The RA-induced differentiated SY5Y cells showed consistently a more slender and longer bipolar morphology than cells which were induced to differentiate whith NGF.

Total RNA was prepared from cells at different stages during differentiation. To quantitate total APP gene transcription we used Northern blot hybridizations. The randomly primed 1182 bp *Bam*HI/*Xho*I APP770 cDNA fragment was hybridized to the filters to assess total APP expression (see Fig. 2 a). Because we were interested in the time course of APP gene induction, several time points of induction of differentiation were chosen for the determination of APP expression. A less than twofold increase in total APP mRNA is seen during the first 8 h of RA administration (Fig. 6). The marked increase in APP gene expression peaks at day 4 in culture with RA, with a induction factor of 10.2. APP mRNA remains elevated through day 23, the latest time point tested, with a factor of 4.3 compared with control cells (Fig. 6). We did not observe signals indicative of APP transcripts of lower molecular weight, given the sensitivity of Northern hybridizations and the quality of our RNA preparations (data not shown).

When we analyzed splicing by measuring quantitavely the ratios of the APP transcripts in the same RNA preparation with probe M8.11 in a S1 nuclease

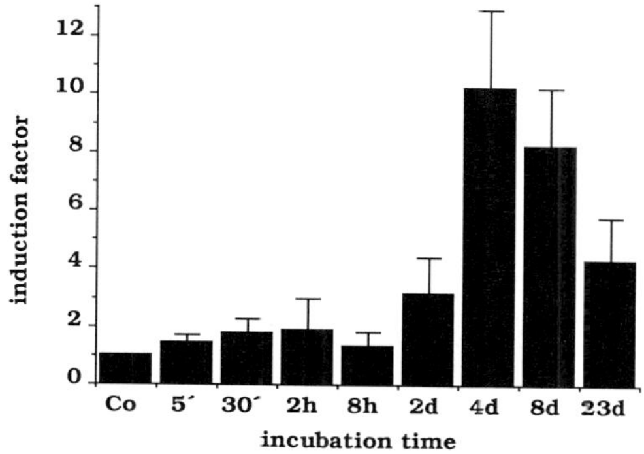

Fig. 6. Induction of total amyloid precursor protein gene expression in SY5Y neuroblastoma cells after differentiation with retinoic acid. Cells were incubated with retinoic acid for the times indicated

protection assay, a significant shift in the splicing pattern was seen within a slightly different time frame as observed for the increase of total APP mRNA. Whereas in control cells grown at a low density the relative amount of APP695 transcript is 54 %, in cells which were treated with RA over a period of 8 days the APP695 transcript makes up 70 % (Fig. 7). In contrast, APP770 mRNA is reduced to 3.5 % and APP751 mRNA, to 26.5 % of total APP mRNA compared with 8.5 % of APP770 mRNA and 37.5 % of APP751 mRNA in undifferentiated cells grown at low density (Fig. 7). This change of the splicing pattern is only seen after an initial maximal increase of total APP mRNA on day 4 and is associated with an increase of total APP expression. The more immediate response of the total APP gene expression of a factor of about 2, which peaks at 2 h after RA-administration, does not lead to a significant change in the splicing pattern (Fig. 7).

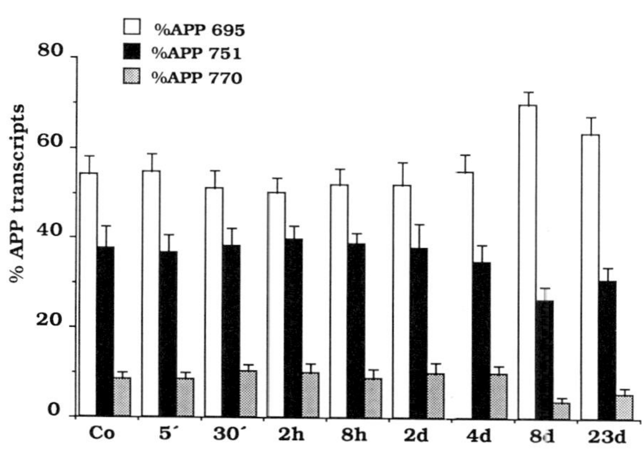

Fig. 7. Relative amyloid precursor protein (*APP*) transcript amount in SY5Y neuroblastoma cells incubated with retinoic acid for the times indicated

When the increase of the total APP gene transcripts of differentiated neuroblastoma cells is expressed as the ratio of each of the three APP transcripts, then the shortest one, APP695, is increased by a factor of 10 on days 4 and 8, the APP751 transcript by a factor of 9.5 on day 4 and 6 on day 8, and the APP770 transcript by a factor of 12 on day 4 and only 3.4 on day 8. These data demonstrate an alternative splicing pattern of the APP hnRNA in differentiated SY5Y cells when induced with RA in favor of APP695 mRNA. The cells show a splicing pattern which is similar to differentiating neurons of a developing human brain in vivo. This change in the splicing ratio is associated with a pronounced increase of total APP transcripts, as is the case for the developing human brain, in which APP expression is much higher than in adult brain [65].

Discussion

In order to study the splicing pattern of the APP gene, we applied a S1 nuclease assay which enabled us to determine the splice ratios of the APP695, APP751 (together with APP563), and APP770 transcripts in a single experiment. Under the assumption of similar half-lives for the known splice forms, the detection of all major APP transcripts in a single sample should allow an accurate ratio determination. Accordingly, by comparing samples of the same age group with different postmortem delays, we would expect a difference in the ratio of the alternative splice forms if the half-lives of the APP transcripts are different. This was not observed and may be interpreted as indicating that the stability of the corresponding splice forms is identical for all transcripts detected in our nuclease protection experiments. This would also be in agreement with the stability of APP transcripts determined in vitro for SY5Y cells (unpublished results). For the cell culture experiments, our protocol should exclude quantification artifacts due to varying mRNA stability. Furthermore, the fragments generated in our nuclease protection experiments correspond to internal mRNA sequences, which are more stable than more proximal and distal ones [6]. This assay was used to determine the splicing pattern in developing as well as in fully developed brain of normal subjects, those with AD, and age-matched controls for the AD patients.

In fetal brain, APP695 mRNA is dominant, making up more than 90 % of the total, whereas APP751 mRNA accounts for approximately 9 % and APP770 mRNA, for less than 1 %.

A totally different situation is observed in differentiated brain of adult individuals, of individuals with pathologically confirmed AD, as well as of the corresponding control cases. In all samples examined, the relative amount of APP751 mRNA increases drastically to 50 % and more, becoming the dominant splice form in the temporal cortex. In contrast, the relative amount of APP695 mRNA decreases in all older subjects, including those with and without AD, to approximately 40 % or less of total APP transcripts. The younger ones aged 31 and 47 years, show approximately equal amounts of APP695 and APP751 mRNA. The relative increase of APP770 mRNA with age is roughly two- to eightfold and thus follows that of APP751 mRNA. This means that the ratios of APP695/APP751 mRNA and APP695/APP770 mRNA change drastically with development in favor of the two longer transcripts. The ratio of APP751/APP770 mRNA does not vary drastically during development and subsequent aging but rather seems to stay relatively constant with minor variations, APP751 mRNA always predominating over APP770 mRNA by a factor of approximately 6 – 10 (see Fig. 3). This is also true for three of the five subjects with AD examined. However, in the other two subjects with AD, the relative amount of APP770 mRNA was found to be elevated by a factor of 2, thus leading to a ratio of APP751/APP770 mRNA of only about 4 (Fig. 3, 82-year and 94-year cases). Because this increase is not a consistent feature of all cases of AD analyzed by us, we conclude that alternative splicing is not the crucial factor in the etiology.

It has already been reported that overall expression of APP mRNA is much higher in a fetal brain than in adult diffentiated brain [64]. Taken together, these results suggest that in the fully developed brain the APP695 transcript is more strongly down-regulated than are the two longer APP transcripts.

Since neurons are the major source of APP transcripts in the brain, the mRNA ratios detected with the nuclease protection assay reflect basically the expression status of neurons in brain [22, 60]. The data obtained from nonneuronal tissues and nonneuronal tumor cell lines clearly show a lower overall expression of the APP gene, with a splicing pattern generally dominated by APP751 mRNA and APP695 mRNA making up less than 15 % [33, 64, 67]. The decrease of APP695 transcripts in the human cortex with age is thus mainly due to a reduction of APP695 transcripts in neurons, supporting the above-mentioned suggestions. It is therefore likely that this shift in the splice ratios obtained in our S1 nuclease assay reflects the change in the splicing program of neurons during development and thereafter.

If one assumes specific *trans*-acting splicing factors to be responsible for the inclusions of exons 7 and 8, giving rise to APP751 and APP770 mRNA, respectively, one could envisage a situation in the developing fetal brain in which these factors are expressed at a low level, resulting in a high proportion of the APP695 transcript [61]. According to this model, neurons increase the amount of specific *trans*-acting splicing factors later in development, thereby generating more of the APP mRNAs including exon 7, and exon 7 – 8 together, with exon 8. Alternatively, a general reduction of APP hnRNA in the adult developed state would also lead to a higher relative amount of the longer APP splice forms, assuming a constant expression of the *trans*-acting splice factors throughout development.

One could also imagine specific *trans*-acting splicing factors expressed at a high level in the developing brain responsible for the exclusion of exons 7 and 8 and giving rise to mainly APP695 mRNA. Down-regulation of these factors later in development and thereafter would result in more APP transcripts including exon 7, and exons 7 and 8 together.

To study the alternative splicing during development, we used a model system, the human neuroblastoma cell line SY5Y, which can be induced to differentiate in cell culture upon exposure to several agents [11, 52]. In undifferentiated cells growing at low density (logarithmic phase) we observed a splice pattern similar to the one seen in an adult brain, with APP695 mRNA being the dominant transcript, accounting for 54 % of total APP mRNA. Cells which were harvested at medium density show a reduced relative amount of APP695 mRNA (49.5 %) and a slightly higher relative amount of APP751 mRNA (40 %) and APP770 mRNA (10.5 %). If we induce differentiation with RA, total APP mRNA increases by a factor of 10.2 after 4 days of treatment. At this time, no significant change in the splicing pattern is observed, since all APP transcripts are produced in proportions similiar to those in control cells. Total numbers of APP transcripts remain high throughout day 8, and the neuroblastoma cells show the phenotype of differentiated neurons, with the typical bipolar shape and a network of connected processes. Associated with these morphological changes is a shift of the splicing pattern. If we compare the APP expression of these cells with that seen in fetal brain, it appears that the SY5Y cells switched to a more developmentally specific gene expression. In these RA-induced differentiated SY5Y cells, APP695 mRNA accounts for 70 % of total APP mRNA compared with 54 % in low-density control cells. The amount of APP695 mRNA is 10.6 times higher compared with control cells. APP751 mRNA is induced by a factor of 5.8 and APP770 mRNA, by a factor of 3.4. This significant change in the splicing pattern in favor of the APP695 transcript is associated with a sprouting of the neuroblastoma cells in culture [41]. It is interesting to note that these cells closely resemble developing neurons in fetal brain, which establish synaptic contacts. A large quantity of the protease inhibitor containing forms, APP751 and APP770, would not suffice to supply the need of a high protease activity to allow such a cellular plasticity in the developing brain. The phenotype of differentiated neuroblastoma cells could be maintained throughout the 2nd week; after the 3rd week a considerable amount of nondifferentiated, flattened cells overgrew the network of differentiated cells, significantly increasing the cell density. The decreasing relative amount of APP695 mRNA of 63.5 % (induction factor of 5) and a generally lower APP gene expression (induction factor of 4.3) on day 23 of RA treatment supports the observed heterogenous phenotype.

When the neuroblastoma cells were induced with NGF, the increase of total APP transcripts was fivefold. This is not associated with a change in the splicing pattern, even after 8 days of treatment with NGF. The neurotrophic factor NGF is known to be important in the maintenance and survival of sympathetic and sensory neurons of the peripheral nervous system [3]. It has been shown that NGF stimulates the outgrowth of sympathetic neurons into the CNS [39]. It is capable of inducing differentiation of the pheochromocytoma cell line PC 12 into cells of the phenotype of typical primary ganglia in cell culture [26]. RA, however, is a potent morphogen which plays an important role in the development of the CNS of vertebrates [15]. It has been shown to be involved in the establishment of an anteroposterior axis in the CNS and during limb development in vertebrates [14, 17]. It binds to nuclear receptors (hRAR) which are ligand-dependent transcription factors, thereby activating their DNA binding capability and regulating gene expression through responsive elements in the promoter region [19, 54]. The RA receptors are structurally related to glucocorticoid receptors [18]. So far, three different receptors for RA, termed α, β, and γ, have been described. They are expressed in a tissue-specific manner and show different affinities for the ligand [14].

Neuroblastoma cells differentiated with RA consistently exhibit a different morphological phenotype from cells differentiated with NGF and later also grew to a slightly higher density. A possible explanation for the differences observed in APP expression in RA-induced differentiated cells could be a regulation of genes involved in the expression of specific splicing factors responsible for the alternative splicing of the APP gene [27]. These RA target genes would not be affected by NGF, thus not leading to a change in the splicing program in NGF-treated cells.

We conclude that RA treatment of cultured SY5Y cells leads to changes that vary both in quantitative and qualitative terms when compared with cells treated with NGF. First, the up-regulation of total APP expression is more pronounced with RA than with NGF. Second, RA exerts its effects on the APP gene after approximately 4 days, whereas NGF treatment peaks only after about 8 days. Third, RA has a significant effect on alternative splicing of the APP mRNA, switching to a more fetal brain specific splicing pattern. This change is not directly correlated with up-regulation of the APP expression. These lines of evidence provide insight essential for designing new approaches that will ultimately delineate alternative splicing of the APP gene transcripts and provide clues that might help to unravel the function of the APP forms.

References

1. Alzheimer A (1907) Über eine eigenartige Erkrankung der Hirnrinde. Allg Z Psychiatr 64: 146 – 148
2. Ausubel FM, Brent R, Kingston RE, Moore DD, Seidman JG, Smith JA, Struhl K (1989) (ed) Current protocols in molecular biology. Wiley, New York
3. Barde YA, Edgar D, Thoenen H (1983) New neurotrophic factors. Ann Rev Physiol 45: 601 – 612
4. Basler K, Oesch B, Scott M, Westaway D, Wälchli M, Groth DF, McKinley MP, Prusiner SB, Weissmann C (1986) Scrapie and cellular PrP isoforms are encoded by the same chromosomal gene. Cell 46: 417 – 428
5. Berk AJ, Sharp PA (1977) Sizing and mapping of early adenovirus mRNA's by gel electrophoresis of S1 endonuclease-digested hybrids. Cell 12: 721 – 728
6. Brawerman G (1987) Determinants of messenger RNA stability. Cell 48: 5 – 6
7. Brion JP, Cheetham ME, Robinson PA, Couck AM, Anderton BH (1987) Isolation of cDNA's coding for epitopes shared by microtubule-associated proteins and neurofibrillary tangle in Alzheimer's disease. FEBS Lett 21: 28 – 32

8. Chirgwin JD, Przybla AE, Mac Donald RJ, Rutter WJ (1979) Isolation of biologically active ribonucleic acid from sources enriched in ribonuclease. Biochemistry 18: 5294–5299
9. Clark L, Carbon J (1976) MRC OX-2 antigen: a lymphoid/neuronal membrane glycoprotein with a structure like a single immunoglobulin light chain. Cell 9: 91
10. Cohen ML, Golde TE, Usiak MF, Younkin HH, Younkin SG (1988) In situ hybridisation of nucleus basalis neurons show increased β-amyloid mRNA in Alzheimer's disease. Proc Natl Acad Sci USA 85: 1227–1237
11. Cole GM, Wu K, Timiras PS (1985) A culture model for age-related human neurofibrillary pathology. Int J Devl Neurosci 3: (1), 23–32
12. Dammerman M, Goldstein M, Yen SH, Shafit-Zagardo B (1988) Isolation and characterization of cDNA clones encoding epitopes shared with Alzheimer neurofibrillary tangles. J Neurosci Res 19: 43–51
13. De Sauvage F, Octave JN (1989) A novel mRNA of the A4 amyloid precursor gene coding for a possible secreted protein. Science 245: 651–653
14. Dolle P, Ruberte E, Kastner P, Petkovich M, Stoner CM, Gudas LJ, Chambon P (1989) Differential expression of genes encoding a, b, and g retinoic acid receptors and CRABP in the developing limbs of the mouse. Nature 342: 702–705
15. Durston AJ, Timmermans JPM, Hage WJ, Hendriks HFJ, de Vries NJ, Heideveld M, Nieuwkoop PD (1989) Retinoic acid causes an anteroposterior transformation in the developing central nervous cystem. Nature 340: 140–144
16. Dyrks T, Weidemann A, Multhaup G, Salbaum JM, Lemaire HG, Kang J, Müller-Hill B, Masters CL, Beyreuther K (1988) Identification, transmembrane orientation and biogenesis of the amyloid A4 precursor of Alzheimer's disease. EMBO J 7: 949–957
17. Eichele G (1989) Retinoids and vertebrate limb pattern formation. Trends Genet 5: 246–251
18. Evans RM (1988) The steroid and thyroid hormone receptor super-family. Science 240: 889–895
19. Giguere V, Ong ES, Segui P, and Evans RM (1987) Identification of a receptor for the morphogen retinoic acid. Nature 330: 624–629
20. Glenner GG (1988) Alzheimer's disease: its proteins and genes. Cell 52: 307–308
21. Glenner GG, Wong CW (1984a) Alzheimer's disease: initial report of the purification and characterization of a novel cerebrovascular amyloid peptide. Biochem Biophys Res Comm 120: 885–890
22. Goedert M (1987) Neuronal localization of amyloid beta protein precursor mRNA in normal human brain and in Alzheimer's disease. EMBO J 6: 3627–3632
23. Goedert M, Wischik CM, Crowther RA, Walker JE and Klug A (1988) Cloning and sequencing of the cDNA encoding a core protein of the paired helical filament of Alzheimer disease: identification as the microtubule-associated protein tau. Proc Natl Acad Sci USA 85: 4051–4055
24. Golde TE, Estus S, Usiak M, Younkin LH, Younkin SG (1990) Expression of β-amyloid protein precursor mRNA's: recognition of a novel alternatively spliced form and quantification in Alzheimer's disease using PCR. Neuron 4: 253–267
25. Goldgaber D, Lerman MI, McBride OW, Saffiotti U, Gajdusek DC (1987) Characterization and chromosomal localization of a cDNA encoding brain amyloid of Alzheimer's disease. Science 235: 877–880
26. Green LA, Tischler AS (1976) Establishment of a noradrenergic-clonal line of rat adrenal pheochromocytoma cells which respond to nerve growth factor. Proc Natl Acad Sci USA 73: 2424–2428
27. Green S, Chambon P (1988) Nuclear receptors enhance our understanding of transcription regulation. Trends Genet 4: 309–314
28. Higgins GA, Lewis DA, Bahmanyar S, Goldgaber D, Gajdusek DC, Young WG, Morrison JH, Wilson MC (1988) Differential regulation of amyloid β-protein mRNA expression within hippocampal neuronal subpopulations in Alzheimer disease. Proc Natl Acad Sci USA 85: 1297–1301

29. Johnson SA, Rogers J, Finch CE (1989) APP-695 transcript prevalence is selectively reduced during Alzheimer's disease in cortex and hippocampus but not in cerebellum. Neurobiol Aging 10: 267 – 272
30. Kang J, Müller-Hill B (1990) Differential splicing of Alzheimer's disease amyloid A4 precursor RNA in rat tissues: preA4$_{695}$ mRNA is predominantly produced in rat and human brain. Biochem Biophys Res Commun 166: 1192 – 1200
31. Kang J, Lemaire HG, Unterbeck A, Salbaum JM, Masters CL, Grzeschik KH, Multhaup G, Beyreuther K, Müller-Hill B (1987) The precursor of Alzheimer's disease amyloid A4 protein resembles a cell-surface receptor. Nature 325: 733 – 736
32. Katzman R (ed) (1983) Biological aspects of Alzheimer disease. Banbury Report 15. Cold Spring Harbor Laboratory, New York
33. Kitaguchi N, Takahashi Y, Tokushima Y, Shirohiri S, Ito H (1988) Novel precursor of Alzheimer's disease amyloid protein shows protease inhibitory activity. Nature 331: 530 – 532
34. Knauer DJ, Cunningham DD (1982) Epidermal growth factor carrier protein binds to cells via a complex with released carrier protein nexin. Proc Natl Acad Sci USA 79: 2310 – 2314
35. König G, Masters CL, Beyreuther K (1990a) Retinoic acid induced differentiated neuroblastoma cells show increased expression of the βA4 amyloid gene of Alzheimer's disease and an altered splicing pattern. FEBS Lett 269: 305 – 310
36. König G, Salbaum JM, Wiestler O, Lang W, Schmidt HP, Masters CL, Beyreuther K (1991) Alternative splicing of the βA4 amyloid gene of Alzheimer's disease in cortex of control and Alzheimer's disease patients. Mol Brain Res 9: 259 – 262
37. Koo EH, Sisodia SS, Cork LC, Unterbeck A, Bayney RM, Price DL (1990) Differential expression of amyloid precursor protein mRNA's in cases of Alzheimer's disease and in aged nonhuman primates. Neuron 2: 97 – 104
38. Lemaire HG, Salbaum JM, Multhaup G, Kang J, Bayney RM, Unterbeck A, Beyreuther K, Müller-Hill B (1989) The preA4$_{695}$ precursor protein of Alzheimer's disease A4 amyloid is encoded by 16 exons. Nucl Acid Res 17: 517 – 522
39. Levi-Montalcini R (1975) Worden FG, Swazey JP, Adelman G (eds) Neuroscience: paths of discovery. MIT Press, Cambridge, pp 245 – 265
40. Lewis DA, Higgins GA, Young WG, Goldgaber D, Gajdusek DC, Wilson MC, Morrison JH (1988) Distribution of precursor amyloid β-protein messenger RNA in human cerebral cortex: relationship to neurofibrillary tangles and neuritic plaques. Proc Natl Acad Sci USA 85: 1691 – 1695
41. Machida CM, Rodland KD, Matrisian L, Magun BE, Ciment G (1989) NGF induction of the gene encoding the protease transin accompanies neuronal differentiation in PC12 cells. Neuron 2: 1587 – 1596
42. Masters CL, Beyreuther K (1988) Amyloidogenic A4 protein subunit: clues to the neurofibrillary tangle, Alzheimer plaque, and congophilic angiopathy. In: Terry RD (ed) Aging and the brain, vol 32. Raven, New York, pp 183 – 204
43. Masters CL, Multhaup G, Simms G, Pottgiesser J, Martins RN, Beyreuther K (1985) Neuronal origin of a cerebral amyloid: neurofibrillary tangles of Alzheimer's disease contain the same protein as the amyloid of plaque core and blood vessels. EBO J 4: 2757 – 2763
44. Masters CL, Simms G, Weinman NA, Multhaup G, McDonald BL, Beyreuther K (1985) Amyloid plaque core protein in Alzheimer disease and Down syndrome. Proc Natl Acad Sci USA 82: 8245 – 4249
45. Mobley WC, Neve RL, Prusiner SB, McKinley MP (1988) Nerve growth factor increases mRNA levels for the prion protein and the β-amyloid protein precursor in developing hamster brain. Proc Natl Acad Sci USA 85: 9811 – 9815
46. Mori H, Kondo J, Ihara Y (1987) Ubiquitin is a component of paired helical filaments in Alzheimer's disease. Science 235: 1641 – 1644
47. Müller-Hill B, Beyreuther K (1989) Molecular biology of Alzheimer's disease. Annu Rev Biochem 58: 287 – 307

48. Neve RL, Finch EA, Dawes LR (1988) Expression of the Alzheimer precursor gene transcript in the human brain. Neuron 1: 669 – 677
49. Oltersdorf T, Fritz LC, Schenk DB, Lieberburg I, Johnson-Wood KL, Beattie EC, Ward PJ, Blacher RW, Dovey HF, Sinha S (1989) The secreted form of the Alzheimer's amyloid precursor protein with the Kunitz domain is protease nexin-II. Nature 341: 144 – 147
50. Pahlman S, Ruusala A, Abrahamson L, Mattson MEK, Esscher T (1984) Retinoic acid-induced differentiation of cultured human neuroblastoma cells: a comparison with phorbolester-induced differentiation. Cell Differ 14: 135 – 144
51. Palmert MR, Golde TE, Cohen ML, Kovacs DM, Tanzi RE, Gusella JF, Usiak MF, Younkin LH, Younkin SG (1988) Amyloid protein precursor messenger RNA's: differential expression in Alzheimer's disease. Science 241: 1080 – 1084
52. Pennypacker KR, Kuhn DM, Billingsley ML (1989) Changes in expression of tyrosine hydroxylase immunoreactivity in human SMS-KCNR neuroblastoma following retinoic acid or phorbol ester-induced differentiation. Mol Brain Res 5: 251 – 258
53. Perez-Polo JR, Wehrbach-Perez K, Tiffany-Castiglioni E (1979) A human clonal cell line modell of differentiating neurons. Dev Biol 71: 341 – 355
54. Petkovich M, Brand N, Krust A, Chambon P (1987) A human retinoic acid receptor which belongs to the family of nuclear receptors. Nature 330: 444 – 450
55. Ponte P, Gonzalez-DeWhitt P, Schilling J, Miller J, Hsu D, Greenberg B, Davis K, Wallace W, Lieberburg I, Fuller F, Cordell B (988) A new A4 amyloid mRNA contains a domain homologous to serine protease inhibitors. Nature 331: 525 – 527
56. Robakis NK, Ramakrishna N, Wolfe G, Wisniewski HM (1987) Molecular cloning and characterization of a cDNA encoding the cerebrovascular and neuritic plaque amyloid peptides. Proc Natl Acad Sci USA 84: 4190 – 4194
57. Ross RA, Spengler BA, Biedler JL (1983) Coordinate morphological and biochemical interconversion of human neuroblastoma cells. JNCI 71: 741 – 747
58. Saitoh T, Sundsmo M, Roch JM, Kimura N, Cole G, Schubert D, Oltersdorf T, Schenk DB (1989) Secreted form of amyloid precursor is involved in the growth regulation of fibroblasts. Cell 58: 615 – 622
59. Salbaum JM, Weidemann A, Lemaire HG, Masters CL, Beyreuther K (1988) The promoter of Alzheimer's disease amyloid A4 precursor gene. EMBO J 7: 2807 – 2813
60. Shivers B, Hilbich C, Multhaup G, Salbaum JM, Beyreuther K, Seeburg P (1988) Alzheimer's disease amyloidogenic glycoprotein: expression pattern in rat brain suggests role in cell contact. EMBO J 7: 1365 – 1370
61. Streuli M, Saito H (1989) Regulation of tissue-specific alternative splicing: exon-specific *cis*-elements govern the splicing of leukocyte common antigen pre-mRNA. EMBO J 8: 787 – 796
62. Tanaka S, Nakamura S, Ueda K, Kameyama M, Shiojiri S, Takahashi Y, Kitaguchi N, Ito H (1988) Three types of amyloid protein precursor mRNA in human brain: their differential expression in Alzheimer's disease. Biochem Biophys Res Commun 157 (2): 472 – 479
63. Tanzi RE, Gusella JF, Watkins PC, Bruns GAP, St George-Hyslop P, van Keuren ML, Patterson D, Pagan S, Kurnit DM, Neve RL (1987) Amyloid β-protein gene: cDNA, mRNA distribution, and genetic linkage near the Alzheimer locus. Science 235: 880 – 884
64. Tanzi RE, McClatchey AI, Lamperti ED, Villa-Komaroff L, Gusella JF, Neve RL (1988) Protease inhibitor domain encoded by an amyloid protein precursor mRNA associated with Alzheimer's disease. Nature 331: 528 – 530
65. Thomas P (1980) Hybridization of denatured RNA and small DNA fragments transferred to nitrocellulose. Proc Natl Acad Sci USA 77: 5201 – 5205
66. Tsuchiya Y, Hattori M, Hayashida K, Ishibashi H, Okubo H, Sakaki Y (1987) Sequence analysis of the putative regulatory region of the rat alpha-2-macroglobulin gene. Gene 57: 73 – 80

67. Weidemann A, König G, Bunke D, Fischer P, Salbaum JM, Masters CL, Beyreuther K (1989) Identification, biogenesis, and localization of precursors of Alzheimer's disease A4 amyloid protein. Cell 57: 115 – 126
68. Yanish-Perron C, Vieira J, Messing J (1985) Improved M13 phage cloning vectors and host strains: nucleotide sequences of the M13mp18 and pUC19 vectors. Gene 33: 103 – 119

Aging Brain and Hypothalamic-Pituitary-Adrenal Function: Implications for Dementia of the Alzheimer Type

E. Martignoni, A. Costa, F. Petraglia, A.R. Genazzani, and G. Nappi

Introduction

Traditionally, senescence is a period of reduced ability to cope with the tasks of everyday life. The various organs and systems in aged subjects normally work at a sufficient level but fail to respond adequately to any further environmental demand [1]. Age primarily affects brain function: impaired cognition represents a common feature of late life, readily encountered in clinical practice.

From a nosographic standpoint, the aging brain appears as a spectrum of multifactorial disorders, and it seems difficult to classify them using a clinical/descriptive approach. Indeed, the various changes in behavior, cognition, and performance appear among different clinical forms, ranging from the so-called chronic organic brain syndrome [2] to clearly pathological conditions such as the dementias belonging to the Alzheimer group (DAT).

Strong evidence supports a derangement of various hypothalamic-pituitary axes during aging. These aspects have been tentatively explained either in terms of defects of hormone degradation/secretion or of an impaired sensitivity of target organs to the effects of hormone. Changes in receptor affinity/number or in postreceptorial responses have been suggested as explanations of the impaired function of the endocrine systems [3]. Modifications occurring in the hypothalamus-pituitary-adrenocortical (HPA) axis have roused great interest in the past few years. Indeed, age seems to be directly correlated with basal plasma cortisol concentrations [4] and dexamethasone resistance [5], thus suggesting hyperactivity of the HPA axis.

In DAT patients, several studies have indicated remarkable changes in the HPA axis activity. In particular, the corticotropin releasing factor (CRF)-like immunoreactivity has been recently found to be decreased in the cerebral cortex of patients with histologically verified Alzheimer's disease (AD), whereas CRF receptors appeared reciprocally increased [6]. Since CRF is the hypothalamic peptide regulating pituitary-adrenocortical secretion [7], such results strengthened the hypothesis of an involvement of the HPA axis in DAT. Indeed, in DAT patients basal and post-dexamethasone cortisol levels were found to be higher than in age-matched healthy controls [8, 9]. As a result of these observations, hyperfunction of the HPA axis has been proposed as a neuroendocrine marker of the disease. Sapolsky et al. [10] have demonstrated in rats the potential toxicity of glucocorticoids for the hippo-

Reihe der Villa Vigoni, Band 1
Biology of Aging
Zwilling and Balduini (Eds.)

campus, the main target tissue in the brain for the negative feedback of adrenocortical hormones. The general features of this experimental model seem to be appropriate to humans, in whom corticosterone is replaced by cortisol. The glucocorticoid effect on the hippocampal receptors may reduce the hippocampal inhibitory action on HPA axis activity, thus resulting in an adrenocortical hyperfunction and elevated cortisolemia [11]. We measured the cerebrospinal fluid (CSF) CRF levels in a group of DAT patients; moreover, in order to assess the functional regulation of cortisol secretion, we evaluated the circadian changes in plasma cortisol levels and the cortisol response to two simple challenges, i.e., dexamethasone suppression test (DST) and CRF stimulation test.

Materials and Methods

The study group consisted of 10 inpatients suffering from DAT, whose characteristics are reported in Table 1. All subjects met the DSM III-R criteria for primary degenerative dementia [12]. Dementia severity was assessed by the Mini Mental State (MMS) [13] and scored by means of the Clinical Dementia Rating (CDR) [14] Scales. Eleven volunteers (7 men and 4 women), healthy and age-matched (64.5 ± 6.4 years, range 53 – 70 years), were the controls for plasma cortisol evaluation. As controls for CSF analysis we considered 11 patients (9 men and 2 women), aged 67.3 ± 6.9 years (range 55 – 79 years), who underwent a lumbar puncture for diagnostic purposes. In all of them, any neurodegenerative disease was ruled out. DAT subjects and controls were hospitalized patients, and their investigation started at least 1 week after admission to the neurological department. All the subjects who entered the study were free of any medical illness known to affect cortisol levels or the dexamethasone suppression test (DST) and had been drug-free for at least 2 weeks prior to the examination. According to the Hamilton Rating Scale for Depression (HRSD) [15], patients and controls showed a score <18. Informed consent was obtained from all subjects or from the patient's family.

CSF samples for CRF determination were obtained by lumbar puncture in 9 of the 10 DAT patients and in all controls. Subjects were kept in bed and fasted for 12 h preceding lumbar puncture (performed at 9.00 a.m.). The CSF samples were immediately frozen and stored at – 20 °C until assay.

Table 1. Characteristics of patients with dementia of the Alzheimer type

Number	10
Sex (M:F)	8:2
Age (years), mean (range)	63.8 (48 – 78)
Symptom duration (years), mean (range)	2.4 (1 – 7)
MMS score, mean (range)	16 (7 – 23)
CDR score, range	1 – 2

MMS, Mini Mental State Score; CDR, Clinical Dementia Rating Scale

In order to study the plasma cortisol circadian rhythm, catheters were inserted at 7.30 a.m. after an overnight fast, and heparinized blood samples were obtained at 8.00 a.m. and then at 4-h intervals until 8.00 a.m. of the following day. Plasma was separated, immediately frozen, and kept at −20 °C until assay.

In all subjects, a dexamethasone suppression test was performed by administering orally dexamethasone 1 mg at 11.00 p.m. Blood was sampled at 8.00 a.m., 4.00 p.m., and 11.00 p.m. the next day. Any subject showing cortisol values equal to or greater than 50 ng/ml at 4.00 p.m. was defined as a nonsuppressor.

The CRF test [16] was performed at 9.00 a.m. after an overnight fast. The cubital vein was cannulated 1 h before the test to avoid hormonal increase in response to venipuncture stress and kept open with intravenous saline. Human CRF was administered intravenously at a dose of 100 μg; blood samples were taken 15 min and immediately before the CRF injection, then 15, 30, 60, and 120 min thereafter. The CRF determination procedure (radioimmunoassay, RIA) is described elsewhere [17]. Cortisol plasma levels were assayed by a direct RIA utilizing the commercially available materials supplied by RADIM (Rome, Italy).

Results

The CSF CRF levels in DAT patients (462 ± 288 pg/ml, range 156 – 1071 pg/ml) were significantly higher than in controls (158 ± 122 pg/ml, range 38 – 470 pg/ml) ($P<0.01$). Also, they were found to be inversely correlated to the MMS score ($y=20.41\,x-3.73$; $r=-0.53$; $P<0.05$) but not to the score of the CDR scale, nor to age and symptom duration.

Concerning the plasma cortisol concentrations over 24 h, DAT patients displayed significantly higher values ($P<0.05$) at 8.00 a.m., on both the first (151.05 ± 52.40 ng/ml in DAT patients vs, 91.84 ± 38.86 ng/ml in controls) and second (171.04 ± 53.84 vs. 106.34 ± 22.13 ng/ml, respectively) days. The cosinor analysis [18] showed in DAT patients a higher value of mesor (106.4 ± 28.7 ng/ml) and amplitude (52.5 ± 8.8 ng/ml) than in controls (70.8 ± 22 ng/ml and 31.5 ± 15.6 ng/ml, respectively) ($P<0.01$), whereas the acrophase of cortisol secretion was similar in the two groups.

With the DST, 6 out of the 10 DAT patients showed post-dexamethasone cortisol levels over the cut-off value (50 ng/ml) at 4.00 p.m., whereas a normal suppression was observed in all but one control. When patients were divided into suppressors (S) and nonsuppressors (N), the pre-dexamethasone cortisol levels of the two groups were not significantly different (151.55 ± 34.47 and 160.25 ± 73.97 ng/ml, respectively). Moreover, no difference between S and N was found for mean age and duration of illness.

Patients and controls exhibited a similar plasma cortisol response to CRF administration. They only differed in that the peak values were reached earlier in DAT patients (at 30 min, $P<0.05$ vs. zero values) than in controls (at

Table 2. Summary of the results on hypothalamus-pituitary-adrenocortical axis activity in patients with dementia of Alzheimer type compared with age-matched controls

Plasma cortisol levels	Enhanced (08.00 hours)
Circadian rhythm	Detectable
Acrophase	Unchanged
Mesor	Greater
Amplitude	Greater
DST	More frequent escape
CRF test	Earlier cortisol peak
CSF CRF levels	Increased

DST, dexamethasone suppression test; CRF, corticotropin releasing factor; CSF, cerebrospinal fluid

60 min, $P<0.01$ vs. zero values). The cortisol secretory area did not differ significantly in the two groups (3385.40 ± 7350.37 ng × 120 min/ml for DAT patients and 6590.19 ± 5575.3 ng × 120 min/ml for controls). An inverse correlation was found in the DAT group between age and the maximum percentage increase of cortisol plasma levels in response to CRF ($r = -0.76$; $P<0.01$).

All the results of the study are summarized in Table 2.

Discussion

Previous studies have suggested that hyperactivity of the HPA axis is a biological feature of a large number of DAT patients [8, 9]. A disruption of the negative feedback mechanism of glucocorticoids on hypothalamic CRF release can be proposed as a possible explanation [11]. Our DAT patients displayed morning (8.00 a.m.) plasma cortisol levels which were significantly higher than in controls, along with a lack of dexamethasone-induced suppression in most cases. These findings are in agreement with previous observations which suggest a dysregulation of the HPA axis in DAT. Moreover, the chronobiological evaluation of the cortisol circadian rhythm showed in these patients higher values of mesor and amplitude of oscillations; thus, these data also seem consistent with a poor stability of plasma cortisol levels, which may be easily influenced by even weak environmental stimuli. In the evaluation of the disturbance of this biological rhythm, a decrease in volume and cell number in the human suprachiasmatic nucleus [19] should be taken into consideration.

CSF CRF levels were higher in the DAT group than in controls. This finding is in keeping with the fact that in DAT patients the CRF-like immunoreactivity has been found to be markedly increased in the paraventricular nucleus (PVN) of the hypothalamus [20]. Because most brain CRF is synthesized in the PVN [21] and is probably released from there directly into the CSF [22], the increased CSF CRF content in our DAT patients may reflect the hypothalamic CRF hypersecretion. The CSF CRF increase was found to correlate with the impairment of cognitive functions, as assessed by the MMS. We therefore believe that CSF CRF measurement in these patients could have some clinical meaning, in particular, providing a better tool to follow the progression of dementia. In this light, it may be of interest to evaluate CSF CRF levels in groups of subjects at risk of developing dementia syndromes who do not yet fit the diagnostic criteria for DAT.

The acute cortisol response to CRF in our DAT patients was similar to that in controls, thus suggesting a normal responsiveness of pituitary corticotrophs to the hypothalamic releasing factor. The finding of an earlier peak of cortisol response in DAT (at 30 min) seems rather difficult to explain; however, it may be viewed as a massive discharge of adrenocorticotropin from the pituitary and/or of cortisol from the adrenal cortex following CRF administration, which may represent a challenge to a hyperresponsive system. On the other hand, in the group of DAT patients, we observed an inverse correlation between age and the maximum percentage CRF-induced increase in plasma cortisol levels. This age-related blunted response in DAT could be ascribed to a progressively reduced sensitivity of the pituitary gland to exogenous CRF. Indeed, previous experimental evidence indicates that prolonged exposure to CRF can induce pituitary desensitization with a lower responsiveness to the secretory stimulus [23, 24]. However, to draw any reliable conclusion, a larger population of subjects should be studied.

In conclusion, the evidence of high CSF CRF and "basal" plasma cortisol levels and the abnormal response to dexamethasone suggest central and peripheral increased activity of the HPA axis in DAT. Whether this is due to the lack of hippocampal inhibitory action on hypothalamic CRF release deserves further investigation in humans. However, also on the basis of previous evidence, it can be proposed that, among other neuroendocrine abnormalities, the adrenocortical hyperfunction plays an outstanding role in the evolution of degenerative dementia.

References

1. Finch CE (1979) Neuroendocrine mechanisms and aging. Federation Proc 38: 178 – 183
2. Lehmann HE (1983) Psychopharmacological approaches to the organic brain syndrome. Compr Psychiatry 24: 412 – 420
3. Everett AV, Burgess JA (eds) (1976) Hypothalamus, pituitary and aging. Thomas, Springfield, IL, USA
4. Dilman VM, Lapin IP, Oxenkrug GF (1979) Serotonin and aging. In: Essman WB (ed) Serotonin in health and disease. Spectrum, New York, pp 79 – 86
5. Oxenkrug GF, Pomara N, McIntyre LM, Branconnier RJ, Stanley M, Gershon S (1983) Aging and cortisol resistance to suppression by dexamethasone: a positive correlation. Psychiatry Res 10: 125 – 130
6. De Souza EB, Whitehouse PJ, Kuhar MJ, Price DL, Vale WW (1986) Reciprocal changes in corticotropin-releasing factor (CRF)-like immunoreactivity and CRF receptors in cerebral cortex of Alzheimer's disease. Nature 319: 593 – 595

7. Vale WW, Spiess J, Rivier C, Rivier J (1981) Characterization of a 41-residue ovine hypothalamic peptide that stimulates secretion of corticotropin and B-endorphin. Science 213: 1394 – 1397
8. Davis KL, Davis BM, Greenwald BS, Mohs RC, Mathe AA, Johns CA, Horvath TB (1986) Cortisol and Alzheimer's disease: I. Basal studies. Am J Psychiatry 143: 442 – 446
9. Greenwald BS, Mathe AA, Mohs RC, Levy MI, Johns CA, Davis KL (1986) Cortisol and Alzheimer's disease: II. Dexamethasone suppression, dementia severity, and affective symptoms. Am J Psychiatry 143: 442 – 446
10. Sapolsky RM (1986) Glucocorticoid toxicity in the hippocampus: temporal aspects of synergy with kainic acid. Neuroendocrinology 43: 440 – 444
11. Sapolsky RM, Krey LC, McEwen BS (1986) The neuroendocrinology of stress and aging: the glucocorticoid cascade hypothesis. Endocr Rev 7 (3): 284 – 301
12. American Psychiatric Association (1987) Diagnostic and statistical manual of mental disorders, 3rd ed. revised. APA, Washington DC
13. Folstein M, Folstein S, McHugh P (1975) Mini Mental State. J Psychiatry Res 12: 189 – 198
14. Hughes CP, Berg L, Danzinger WL, Coben LA, Martin RL (1982) A new clinical scale for the staging of dementia. Br J Psychiatry 140: 566 – 572
15. Hamilton M (1960) A rating scale for depression. J Neurol Neurosurg Psychiatry 23: 56 – 62
16. Chrousos GP, Schulte HM, Oldfield EH, Gold PW, Cutler GB Jr, Loriaux DL (1984) The corticotropin releasing factor stimulation test. N Engl J Med 310: 622 – 626
17. Martignoni E, Petraglia F, Costa A, Bono G, Genazzani AR, Nappi G (1990) Dementia of the Alzheimer type and hypothalamus-pituitary-adrenal axis: changes in cerebrospinal fluid corticotropin releasing factor and plasma cortisol levels. Acta Neurol Scand 81: 452 – 456
18. Halberg et al. (1977) Chronobiologia 4 [Suppl. 1]: 1 – 189
19. Swaab DF, Fliers E, Partiman TS (1985) The suprachiasmatic nucleus of the human brain in relation to sex, age and senile dementia. Brain Res 342: 37 – 44
20. Powers RE, Walker LC, De Souza EB, Vale WW, Struble RG Whitehouse PJ, Price DL (1987) Immunochemical study of neurons containing corticotropin releasing factor in Alzheimer's disease. Synapse 1: 405 – 410
21. Vale W, Rivier C, Brown M, Spiess J, Koob G, Swanson L, Bilezikjian L, Bloom F, Rivier J (1983) Chemical and biological characterization of corticotropin releasing factor. Recent Prog Horm Res 39: 245 – 255
22. Berrettini WH, Nurnberger JI, Gold PW, Cheretien M, Chrousos GP, Chan JSD, Goldin LR, Gershon ES (1985) Neuropeptides in human cerebrospinal fluid. Life Sci 37: 1265 – 1270
23. Reisine TD, Hoffman A (1983) Desensitization of corticotropin releasing factor receptors. Biochem Biophys Res Commun 111: 919 – 925
24. Rivier C, Vale W (1983) Influence of the frequency of ovine corticotropin releasing factor administration on adenocorticotropin and corticosterone secretion in rat. Endocrinology 113: 1422 – 1426

Involvement of Proteoglycan Metabolism Regulation in Aging and Atherosclerosis*

G. De Luca and A. Passi

Structure and Functions of Proteoglycans

Proteoglycans (PGs) form a complex and heterogeneous family of macromolecules, which are found in all mammalian tissues. They are particularly abundant in connective tissues (such as cartilage, skin, blood vessel wall) where the extracellular matrix is large and well developed. In these tissues, PGs represent the most important component of the interfibrillar ground substance and interact with collagen fibrils [1 – 3].

PGs are also found in basement membranes and on the surface of cell membranes: some [heparan sulfate (HS)-PGs] seem to be ubiquitous components of cultured mammalian cells, occurring partly as membrane-associated molecules [4].

PGs consist of a core protein to which one or more glycosaminoglycan (GAG) chains and N- and/or O-linked oligosaccharides are covalently linked. GAGs are the characterizing components of PGs and have an unbranched chain structure containing repeating disaccharide units typical of each GAG [1 – 4]. However, this is only a general definition, and GAGs show variations in structural details when prepared from different sources [2].

The structure of most PGs still has to be determined in detail, but it is clear that they show a wide range of forms. There are also differences in molecular size: M_r ranges from 80×10^3 in the case of small "tadpole" cornea/keratan sulfate (KS)-PG to 2×10^6 in the case of cartilage PG containing both KS and chondroitin sulfate (CS). The GAG composition of PGs varies according to both type and number of the chains linked to the core protein (from 1 – 3 in cornea KS-PG to about 150 in cartilage CS-KS-PG). Moreover, the core proteins are different both in size and in amino acid composition, and the same GAG type may occur in more than one type of PG, being linked to different core proteins.

The complexity of PG structures reflects the complexity of their functions, which depend on the unique properties lent by the presence of GAGs. These long polyanion and hydrophilic chains make PGs highly expanded molecules with raised charge density. These properties permit them to resist compressive forces acting on the tissue and to occupy a large domain in the extracellular

*This work was supported by funds from M.U.R.S.T. and from C.N.R. (Progetto Finalizzato "Invecchiamento" 91 1052), Italy.

Reihe der Villa Vigoni, Band 1
Biology of Aging
Zwilling and Balduini (Eds.)

matrix, excluding other macromolecules but retaining permeability to small molecules (water, nutrients, chemical messengers, hormones). Yet, PGs interact with the fibrous components of the extracellular matrix, thus influencing their organization, and maintain the fluid compartment and the water content of the tissue.

Besides these functions, which are performed in all extracellular matrices, PGs are able to play more specialized roles in some tissues. In arterial wall, for example, PGs show antihemostatic properties and may bind lipoprotein lipase, thus regulating the metabolism of triacylglycerol-rich lipoproteins. Moreover, aorta PGs may interact with plasma lipoproteins and regulate their passage through the vessel wall, playing an important role in atherosclerotic degeneration [5, 6].

PG functions are likely to depend on the type of GAGs linked to the core protein. Available data imply that PGs containing homopolymeric GAG chains (CS and KS) have a space-filling role and are thought to be important in maintaining hydration in the extracellular matrix. PGs containing copolymeric GAG chains [dermatan sulfate (DS) and HS] exhibit stronger properties of interactions with other components of the extracellular matrix. This probably depends on differences in chain conformations and charge distributions, which make copolymers particularly able to interact with other macromolecules [2].

Proteoglycans Are Modified During Aging

PGs undergo important modifications with aging. Their content decreases significantly in many tissues: in skin, for example, the total hexuronate content shows a 70 % reduction from birth to old age [7], and this may explain, at least partially, the dehydration of tissue.

In cartilage [8], the size of PG subunits declines with age. CS chains decrease both in number and in size, and chondroitin-4-sulfate decreases more markedly than chondroitin-6-sulfate. On the contrary, KS chains increase both in number and in size. The number of O-linked oligosaccharide chains per core protein is reduced, and as a consequence of all these modifications PG subunits in old cartilage are smaller in size and less glycosylated and show an increase in protein content relative to GAGs. Moreover, the amino acid composition of the core protein changes, and the decrease of serine content in old PGs [9] seems to be particularly important. The hyaluronic acid binding region appears to be a common structural feature of PGs at all ages, and therefore, the ability to form aggregates is maintained. Similar changes in PG structure were found in other species, and they probably depend on different mechanisms acting during the aging process. A variation in core protein gene expression may occur, but also the activity of posttranslational enzymes involved in glycosylation and sulfation is likely to change. Moreover, PGs were shown to undergo proteolytic modifications after their secretion into the extracellular matrix [8], and these degradative reactions contribute to the changes in PG subunit size.

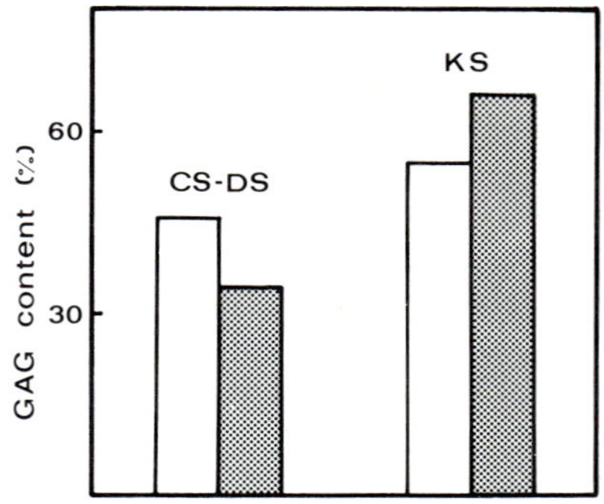

Fig. 1. Percentage content of hexuronate-containing glycosaminoglycans (GAGs) [chondroitin and dermatan sulfates (CS-DS)] and keratan sulfate (KS) in cornea of young calves (▭) and steers (▩). Mean age of young calves was 6 months and steers, 3 years

However, the changes in GAG composition are the best known and the most typical modifications of PGs with aging and contribute to much of the variation in PG structure and function. Both in cartilage and in cornea (Fig. 1) the decrease of CS (containing hexuronate residues) and the concomitant increase of KS (which does not contain hexuronate residues, but galactose) drive the KS/CS ratio to increase with aging, suggesting that hexuronic acid biosynthesis becomes even more difficult in these tissues.

Proteoglycans Are Modified with Atherosclerotic Degeneration

PGs undergo structural and functional modifications with the development of atherosclerotic degeneration. This problem was investigated by studying PGs from human aorta extracted from normal areas and from adjacent atherosclerotic plaques [10].

In all the aortas, total PG content (expressed as hexuronate content) was always lower in atherosclerotic samples than in normal areas of the same artery (Table 1). The decrease was more marked with increasing severity of the

Table 1. Total hexuronate content of normal and affected areas of human thoracic aorta from subjects with different atherosclerotic lesions

	Sex	Age (years)	Normal areas	Affected areas	Percentage variation
a)	♂	71	3.75	3.15	−16.0
b)	♀	74	5.18	4.30	−16.9
c)	♂	61	4.36	3.22	−26.1
d)	♀	49	4.53	3.16	−30.2

Values expressed as mg hexuronate/g lipid-free dry aorta. Aortas (a) and (b) were less markedly affected than aortas (c) and (d)

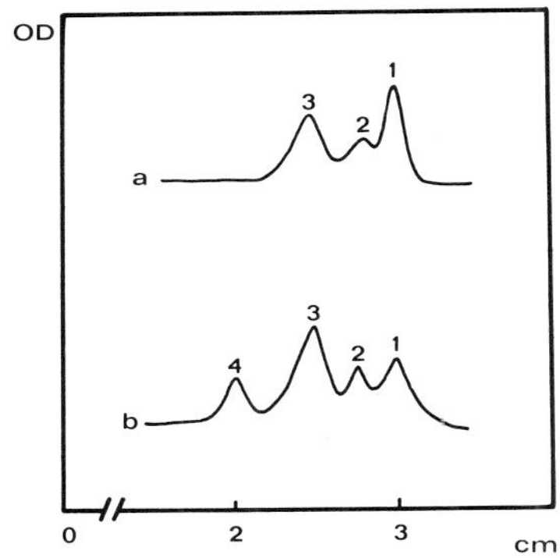

Fig. 2. Photodensitometric (*OD*, optical density) representation of cellulose acetate electrophoresis of glycosaminoglycans (GAGs) in the whole papain digest of normal (*a*) and atherosclerotic areas (*b*) of the same aorta. Band *1* consists of chondroitin sulfate; bands *2* and *4* consist of dermatan sulfate (dermatan sulfate with slower migration may be represented by molecules with a higher degree of iduronation [11], band *3* contains both heparan sulfate and hyaluronic acid (about 70% and 30%, respectively). GAG identification was performed by digestion with chondroitinase ABC (from *Proteus vulgaris*), with chondroitinase AC (From *Arthrobacter aurescens*), with hyaluronidase (from *Streptomyces hyalurolyticus*), and by nitrous acid treatment

disease. It could be interesting to outline that this phenomenon somehow resembled the decrease of PGs occurring with aging in other connective tissues.

The GAG composition of the aortic wall changed: In particular, atherosclerotic samples always contained a higher percentage amount of chondroitinase AC-resistant material, suggesting an increase of the relative content of DS. The electrophoretic separation of GAGs after papain digestion showed that DS with lower mobility was particularly increased in affected areas, and this DS type may be represented by molecules with a higher degree of iduronation [11] (Fig. 2).

The modification of the total GAG composition of the aortic wall suggested that the decrease of PG content in atherosclerotic areas did not apply to all the different types, but only some in particular. The study of intact PGs extracted from the tissue with 4 *M* guanidinium hydrochloride (GuHCl) indicated that this reduction depended chiefly on the marked decrease of the relative content of the largest molecular size PGs. In fact, gel-filtration chromatography of PGs from the arterial wall on Sepharose CL-4B showed that the hexuronate-containing material eluted with the void volume of the column (PGI) was strongly reduced in samples from atherosclerotic minces (Fig. 3). This reduction was more marked with increasing severity of the disease.

PGI contained mostly CS, whereas the material eluted in the inclusion volume of the column (PGII) showed a quite different GAG composition, containing DS and smaller amounts of CS and HS. Moreover, the relative content of DS with slower mobility was higher in PGII from atherosclerotic samples.

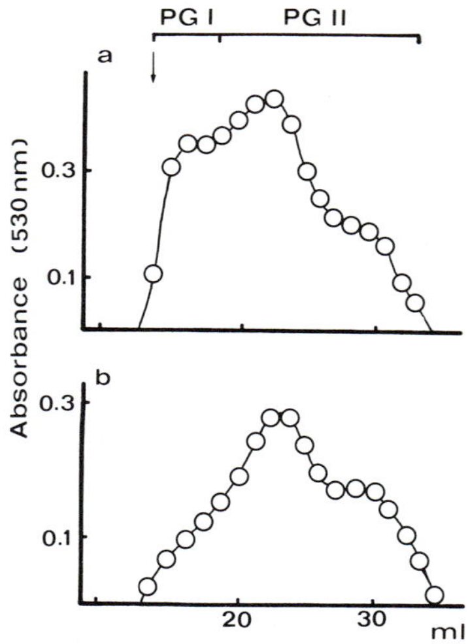

Fig. 3. Sepharose CL-4B chromatography of proteoglycans (PGs) extracted with 4 *M* GuHCl from human aortas. Gel filtration was performed under dissociative conditions (in the presence of 4 *M* GuHCl). Aorta *a* (♀ 47 y) presented with only very slight injuries, whereas aorta *b* (♂ 88 y) showed a very severe and generalized atherosclerotic involvement

It was concluded that atherosclerotic degeneration seemed to be characterized by: (a) a decrease of the total content of PGs; (b) a decrease of the relative content of largest molecular size PGs, containing mostly homopolymeric CS chains; and (c) an increase of the relative content of smaller PGs, rich in copolymeric DS chains, which are more able to interact with other molecules in the extracellular matrix.

Biosynthesis of Proteoglycans and Control Mechanisms During Aging and Atherosclerotic Degeneration

PG modifications occurring with aging and with atherosclerotic degeneration are the results of metabolic changes, which are certainly complex and involve both anabolic and catabolic pathways. We will consider only the biosynthetic pathway of PGs, which may be described as occurring in two main stages: (a) biosynthesis of the core protein and (b) assembly of GAG chains on the core protein, followed by the rapid secretion of PG into the extracellular matrix.

In the case of cultured chondrocytes from a transplantable rat chondrosarcoma [12], the translation of a core protein precursor occurs in rough endoplasmic reticulum, together with N-glycosylation processes and the core protein remains a long time in the cell before the addition of GAG chains.

The assembly of GAG chains on the core protein and the biosynthesis of O-linked oligosaccharides take place in the Golgi apparatus. These later events are very rapid and seem to require only 10% of the total time necessary for the synthesis of the entire PG [12]. The completed PG is then immediately secreted into the extracellular matrix.

In the case of a typical cartilage PG, GAGs comprise about 90% of the total mass, and the assembly of their chains represents the major event in biosynthesis. Taken together with the timing shown for rat chondrosarcoma cells [12], it may be concluded that this biosynthetic stage must be very efficient.

The posttranslational assembly of GAGs is performed by systems of glycosyltransferases, closely located or organised into enzyme complexes within the Golgi apparatus [1, 2]. These transferase systems add a single monosaccharide residue at a time to the nonreducing end of the growing GAG chain, utilizing as a substrate the uridine diphosphate (UDP) derivatives of the monosaccharide units. Therefore, the efficiency of the glycosyltransferase systems depends on the availability of the UDP sugar precursors, which are synthesized by enzymes located in the soluble fraction of the cell [1, 2]. UDP sugars are then transported into the endoplasmic reticulum and Golgi vesicles by an antiport system, introducing them in exchange for UMP [13]. As a consequence, the cytosolic production of sugar nucleotides must be very efficient to support the activity of glycosyltransferase systems in the Golgi apparatus. Moreover, by controlling the type and concentration of the UDP sugars produced in the cytosol, it is possible to induce regulation of the type and amount of GAGs assembled in the Golgi apparatus.

For these reasons, we will consider just this regulatory level, in order to investigate which control mechanisms of GAG precursor biosynthesis could be

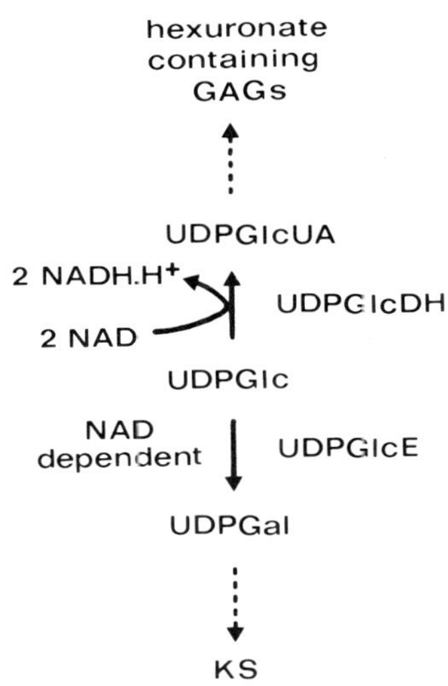

Fig. 4. Uridine diphosphate glucose (*UDPGlc*) utilization in glycosaminoglycan (GAG) precursor biosynthesis. *UDPGlcUA*, UDP glucuronic acid; *NAD*, nicotinamide-adenine dinucleotide; *UDPGlcDH*, UDPGlc dehydrogenase; *UDPGlcE*, UDPGlc-4'-epimerase; *UDPGal*, UDP galactose; *KS*, Keratan sulfate

involved in PG modifications occurring with aging and with atherosclerotic degeneration.

Among the enzymes involved in the production of UDP sugar precursors, a key role is played by those utilizing UDP-glucose (UDPGlc) (Fig. 4). UDPGlc may be oxidized by its dehydrogenase (UDPGlcDH) to UDP-glucuronic acid (UDPGlcUA) for the biosynthesis of CS and other hexuronate-containing GAGs. UDPGlc in turn may be converted into UDP-galactose (UDPGal) by UDP-glucose-4'-epimerase (UDPGlcE) for the biosynthesis of KS, which contains galactose instead of hexuronate residues. Therefore, UDPGlc utilization represents a key branch point in GAG biosynthesis, and its control permits the cell to decide whether hexuronate- or galactose-containing GAGs are to be produced.

Different mechanisms were shown to regulate this biosynthetic step [14 – 21], the most important one being the cytoplasmic UDP xylose concentration.

Role of Uridine Diphosphate Xylose

UDP xylose is produced by decarboxylation of UDPGlcUA and acts as a specific allosteric inhibitor of UDPGlcDH [14 – 16, 18, 21]. On the other hand, it does not affect the activity of UDPGlcE [18]. Therefore, it appears reasonable to hypothesize that the inhibition of UDPGlcDH activity by UDP xylose results in an increase of UDPGlc availability for the epimerase reaction and UDPGal formation.

This hypothesis was tested by incubating an enzyme extract from epiphyseal plate cartilage containing both UDPGlcDH and UDPGlcE with UDPGlc and UDP xylose. Under these experimental conditions, the biosynthesis of UDPGlcUA decreased markedly, whereas the formation of UDPGal increased in agreement with the equilibrium constant of the epimerase reaction, suggesting that this depended on the increased substrate availability for the enzyme (Fig. 5).

The UDP xylose effect on UDPGlc utilization may also influence the relative synthesis of KS and hexuronate-containing GAGs. In fact, Balduini et al. [14] incubated cornea slices with [U-^{14}C]glucose and UDP xylose and showed an increase of KS synthesis concomitant with a decrease of CS formation.

Does this mean that UDPGlcDH activity declines with aging? Histoenzymologic results [22] support this hypothesis, suggesting that UDP xylose may be important for the increase of KS biosynthesis in old cartilage. In hyaline cartilage slices of young dogs, the activity of UDPGlcDH was easily demonstrable, whereas in the same tissue of old dogs, it was nearly absent. Moreover, after incubation of cartilage slices of young animals with UDP xylose, the histoenzymologic demonstration of UDPGlc DH was quite similar to that obtained with old cartilage slices.

These results raise the question of whether or not the cellular concentration of UDP xylose really increases with aging. Some data indicate that it may oc-

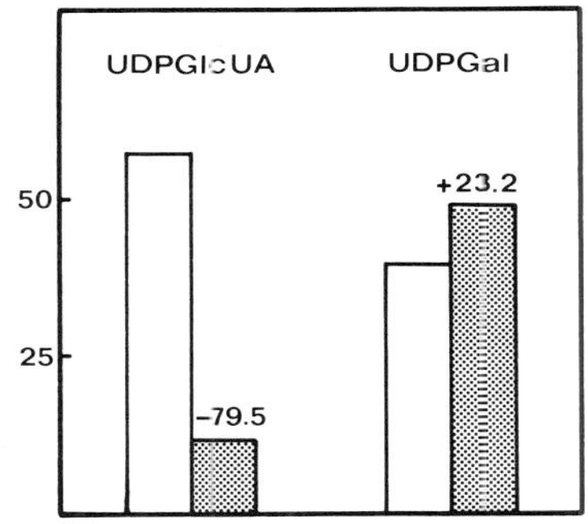

Fig. 5. Uridine diphosphate glucose (UDPGlc) utilization by an enzyme extract from newborn pig epiphyseal plate cartilage containing both UDPGlc dehydrogenase and UDPGlc-4'-epimerase in the absence (▭) and in the presence (▨) of 0.12 m*M* UDP xylose. Values are expressed as nanomoles of UDP glucuronic acid (*UDPGlcUA*) and UDP galactose (*UDPGal*) formed in the reaction mixture after 20 min incubation in the presence of 0.75 m*M* UDPGlc. *On top of the dotted bars* is reported the percentage variation

cur in aging hyaline cartilage. In fact, it was shown that the activity of xylosyltransferase declines with aging in costal cartilage [23] and that the serine content is markedly reduced in the core protein of old cartilage PGs [9].

All hexuronate-containing GAGs are attached to the core protein by means of a xylose residue, which forms an O-glycosidic linkage with serine residues in the core protein [1 – 4]. Therefore, the decrease of both xylosyltransferase activity and serine content in the PG core protein may determine an increase of UDP xylose concentration in the cytosol, inhibiting further synthesis of hexuronate precursors. We may suggest that in young cartilage the xylosyltransferase activity utilizes UDP xylose in the Golgi apparatus for the biosynthesis of the linkage region of hexuronate-containing GAGs. The concentration of UDP xylose in the cytosol remains low, and UDPGlcDH produces UDPGlcUA for the assembly of hexuronate-containing GAG chains. This proposal is confirmed by the finding that the incubation of epiphyseal plate cartilage slices actively synthesizing CS with [U-^{14}C]glucose and UDP xylose did not result in a decrease of CS synthesis nor in an increase of KS production [15].

With aging, the xylosyltransferase activity declines, and fewer serine residues are available in the core protein. Thus, UDP xylose accumulates in the cytosol, inhibiting UDPGlcUA production and synchronizing the biosynthesis of the specific linkage region on the core protein with the production of GAG chain precursors (Fig. 6).

As a consequence, UDPGlc becomes available for the conversion into UDPGal for KS production, which is differently linked to the core protein. In fact, KS is covalently bound to the core protein by an O-glycosidic bond between *N*-acetylgalactosamine and hydroxyl groups of serine and threonine in the case of cartilage and by N-glycosidic bonds between *N*-acetylglucosamine and asparagine in the case of cornea [1 – 3].

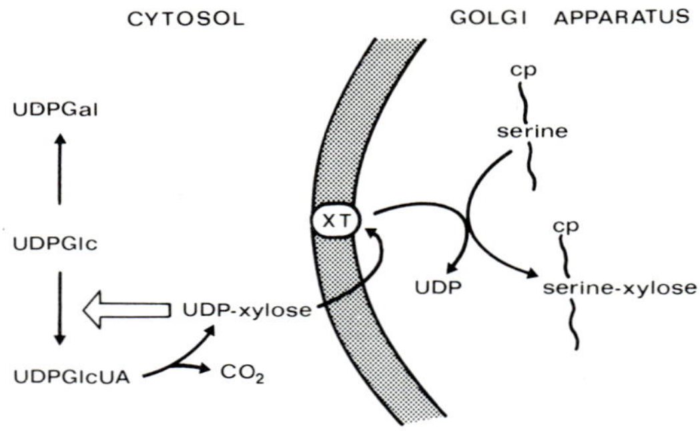

Fig. 6. The regulatory role of uridine diphosphate (UDP) xylose in glycosaminoglycan precursor biosynthesis. *XT*, xylosyltransferase; *cp*, core protein; *UDPGal*, UDP galactose; *UDPGlc*, UDP glucose; *UDPGlcUA*, UDP glucuronic acid

Role of NAD/NADH Ratio

A second mechanism regulating UDPGlc utilization in GAG biosynthesis could be involved in aging. Both oxidation and epimerization reactions in UDPGlc utilization require nicotinamide-adenine dinucleotide (NAD) as a substrate (see Fig. 4).

In the UDPGlcDH reaction, a net production of the reduced form of NAD (NADH) occurs, and the NAD concentration decreases in the cytosol. In the case of UDPGlcE, NAD is released in the oxidized form at the end of the reaction, without decreasing the cytoplasmic NAD concentration. The NAD requirement for both reactions and its different utilization suggest that the relative concentrations of NAD and NADH in the cytosol could selectively affect GAG precursor biosynthesis.

UDPGlcDH is inhibited by NADH. For cartilage and fibroblast enzymes, the inhibition is competitive at low NADH concentrations but shows cooperative allosteric properties with increasing NADH levels [18, 21]. The inhibition is clearly dependent on NAD concentration: When the ratio NAD/NADH is high, UDPGlcDH activity is barely modified, but when the ratio decreases below a critical value (about 15), the production of UDPGlcUA is strongly reduced [18]. Moreover, this regulatory effect was particularly evident at physiological pH [18].

The control mechanism exerted by the NAD/NADH ratio seems to play a particularly important role: It relates the biosynthesis of hexuronic acid precursors to the redox activity of the cell and to the oxygen supply. The production of hexuronate-containing GAGs becomes more difficult with aging, whereas KS production increases both in cartilage and cornea.

Moreover, it must be considered that KS is not widely distributed in mammalian tissues, being present in significant amounts only in cartilage, intervertebral disc, and corneal stroma. These tissues are avascular and obtain their oxygen supply by diffusion from the vessels of the adjoining tissues or from the air, in the case of the cornea.

The KS concentration seems to rise in these tissues with increasing distance from the oxygen supply. In human intervertebral disc and articular cartilages, the greatest amounts of KS are found in the deepest zones [24]. In the cornea, a gradient of KS content has been observed, increasing posteriorly and following the decrease of oxygen tension across the cornea [25 – 27]. Recently, KS as percentage of total GAGs was found to rise with increasing thickness of the corneal stroma [28].

Moreover, KS does not contain hexuronic acids, and the biosynthesis of its precursor, UDPGal, does not require oxidative reactions or oxygen consumption.

For these reasons, KS could be made under conditions of lack of oxygen, while hexuronate-containing GAGs can not.

These data support Scott's "oxygen lack hypothesis" [26, 28], suggesting that KS may be a biological substitute for hexuronate-containing GAGs when the oxygen supply is critical.

In order to test this hypothesis, calf cornea slices were incubated with [U – ^{14}C]glucose at different oxygen tensions. After isolating the radioactive GAGs, the relative amounts of hexuronate-containing GAGs and KS were determined, examining the variations in their ratios with increase of oxygen tension. The percentage radioactivity incorporated in KS consistently increased with the decrease of oxygen tension to the tissue, whereas a concomitant relative decline of the biosynthesis of hexuronate-containing GAGs occurred (Fig. 7). The ratio between total radioactivity incorporated in hexuronate-containing GAGs and that in KS fell with the decrease of oxygen tension in the incubation medium.

The main metabolic mechanism behind these observations may depend on the regulation of the biosynthesis by UDPGlcDH of hexuronate precursors. The UDPGlcDH reaction takes place in the cytosol, lowering the NAD/

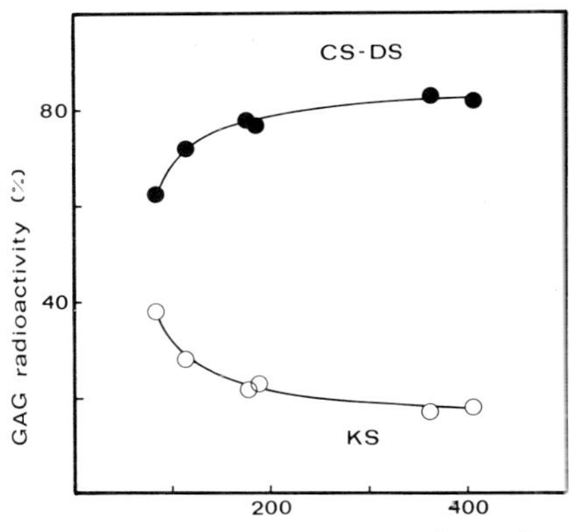

Fig. 7. Percentage glycosaminoglycan (GAG) radioactivity incorporated in hexuronate-containing GAGs [chondroitin and dermatan sulfates (CS-DS)] and in keratan sulfate (KS) produced by corneas incubated with different pO_2

NADH ratio in this compartment. In the presence of an adequate oxygen supply, the mitochondrial shuttle systems support NADH oxidation and permit both a normal glycolytic flow and production of UDPGlcUA for the synthesis of hexuronate-containing GAGs. With oxygen deprivation, UDPGlcDH competes with the glyceraldehyde-3-phosphate dehydrogenase reaction for NAD, reducing the glycolytic rate. The consequent decrease of the NAD/NADH ratio in the cytosol would inhibit further synthesis of UDPGlcUA, leaving greater amounts of UDPGlc available for UDPGal synthesis and KS production. Since the interconversion of UDPGlc into UDPGal does not produce NADH in the cytosol, the KS synthesis affects neither the cytosolic NAD/NADH ratio nor the glycolytic rate. Therefore, the balance of hexuronate-containing GAGs and KS biosynthesis may be determined by oxygen supply to the cell in KS-producing tissues.

It is possible that this mechanism may be working whenever oxygen supply to the tissue falls below a critical value and that it may take place in aging.

The NAD/NADH ratio regulates other enzymes involved in GAG biosynthesis, in particular, NAD-dependent epimerase activities. UDPGlcE from cartilage and UDP-*N*-acetylglucosamine-4'-epimerase from both cartilage and cornea were shown to be affected by this control mechanism [17 – 20].

It is not illogical to suppose that the epimerase catalyzing the conversion of glucuronic acid to iduronic acid may be sensitive to changes in the NAD/NADH ratio. This enzyme does not work on UDP sugar precursors: The C-5 inversion of D-glucuronic acid to L-iduronic acid occurs on the polymer level, by epimerization of D-glucuronic acid residues which have already been transferred to the polysaccharide chain from UDPGlcUA [29, 30]. Therefore, chondroitin D-glucuronosyl-5 epimerase may convert a homopolymeric chain of CS into the copolymeric chain of DS, consequently modifying the properties of interactions of these polysaccharides.

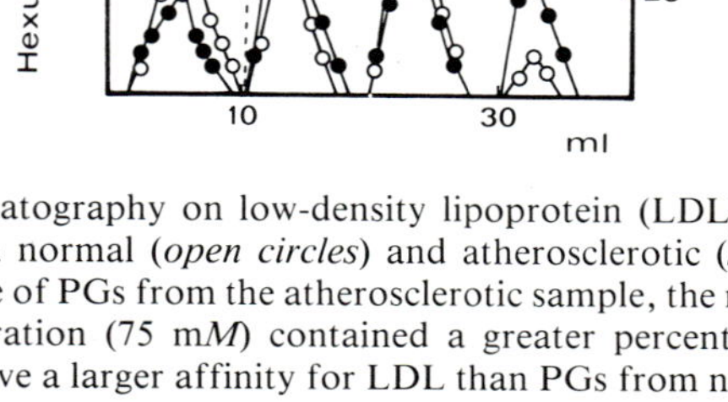

Fig. 8. Affinity chromatography on low-density lipoprotein (LDL)-Sepharose 4B of proteoglycans (PGs) from normal (*open circles*) and atherosclerotic (*solid dots*) areas of the same artery. In the case of PGs from the atherosclerotic sample, the material eluting with the highest NaCl concentration (75 m*M*) contained a greater percentage of hexuronic acid, suggesting that they have a larger affinity for LDL than PGs from normal areas of the same aorta

This represents a very interesting problem, since the DS quantity increases in arterial wall PGs with atherosclerotic degeneration and since PG modifications in affected areas certainly influence the gel sieve properties of the subendothelial extracellular matrix, influencing both the permeability and the retention of lipoproteins.

This aspect was studied by performing affinity chromatography experiments, after coupling human low-density lipoproteins (LDL) to cyanogen bromide (CNBr)-activated Sepharose 4B. Under these experimental conditions, PGs extracted from atherosclerotic samples showed a higher affinity for LDL than those from normal areas of the same aorta (Fig. 8). The demonstration that PGs from atherosclerotic minces contained significantly higher amounts of DS could explain, at least partially, the increase of their interaction properties.

Moreover, the increase of iduronate-containing GAGs in PGs from atherosclerotic areas of human aorta indicates that the GAG biosynthesis control may be significantly involved in the development of atherosclerotic degeneration and in the progressive loss of the normal properties of the arterial wall.

References

1. Handley CJ, Lowther DA, McQuillan DJ (1985) The structure and synthesis of proteoglycans of articular cartilage. Cell Biol Int Rep 9: 753 – 782
2. Hardingham TE (1986) Structure and biosynthesis of proteoglycans. Rheumatology 10: 143 – 183
3. Poole AR (1986) Proteoglycans in health and disease: structures and functions. Biochem J 236: 1 – 14
4. Gallagher JT, Lyon M, Steward WP (1986) Structure and function of heparan sulphate proteoglycans. Biochem J 236: 313 – 325
5. Berenson GS, Radhakrishnamurthy B, Srinivasan SR, Vijayagopal P, Dalferes ER (1985) Proteoglycans and potential mechanisms related to atherosclerosis. Ann NY Acad Sci 454: 69 – 78
6. Wagner WD (1985) Proteoglycan structure and function as related to atherosclerosis. Ann NY Acad Sci 454: 52 – 68
7. Fleischmajer R, Perlish JS, Bashey RI (1973) Aging of human dermis. In: Robert L, Robert B (eds) Frontiers of matrix biology, vol 1. Aging of connective tissues – skin. Karger, Basel, pp 90 – 106
8. Roughley PJ, Mort JS (1986) Aging and the aggregating proteoglycans of human articular cartilage. Clin Sci 71: 337 – 344
9. Roughley PJ, White RJ (1980) Age-related changes in the structure of the proteoglycan subunits from human articular cartilage. J Biol Chem 255: 217 – 224
10. Cherchi GM, Coinu R, Demuro P, Formato M, Sanna G, Tidore M, Tira ME, De Luca G (1990) Structural and functional modifications of human aorta proteoglycans in atherosclerosis. Matrix 10: 362 – 372
11. Cappelletti R, Del Rosso M, Chiarugi VP (1980) A new method for characterization of N-sulphated glycosaminoglycans by a rapid and multisample nitrous acid treatment during an electrophoretic run and its application to the analysis of biological samples. Anal Biochem 105: 430 – 435
12. Kimura JH, Lohmander LS, Hascall VC (1984) Studies on the biosynthesis of cartilage proteoglycans in a model system of cultured chondrocytes from the Swarm rat chondrosarcoma. J Cell Biochem 26: 261 – 278

13. Deutscher SL, Hirschberg CB (1986) Mechanism of glycosylation in the Golgi apparatus. J Biol Chem 261: 96–100
14. Balduini C, Brovelli A, Castellani AA (1970) Biosynthesis of glycosaminoglycans in bovine cornea. The effect of uridine diphosphate xylose. Biochem J 120: 719–723
15. Balduini C, Brovelli A, De Luca G, Galligani L, Castellani AA (1973) Uridine diphosphate glucose dehydrogenase from cornea and epiphysial-plate cartilage. Biochem J 133: 243–249
16. Balduini C, De Luca G, Castellani AA (1989) Biosynthesis of skeletal and corneal keratan sulphate. In: Greiling H, Scott JE (eds) Keratan sulphate. Chemistry, biology, chemical pathology. The Biochemical Society, London, pp 53–64
17. De Luca G, Speziale P, Balduini C, Castellani AA (1975) Biosynthesis of glycosaminoglycans: uridine diphosphate glucose 4'-epimerase from cornea and epiphysial-plate cartilage. Connect Tissue Res 3: 39–47
18. De Luca G, Speziale P, Rindi S, Balduini C, Castellani AA (1976) Effect of some nucleotides on the regulation of glycosaminoglycan biosynthesis. Connect Tissue Res 4: 247–254
19. De Luca G, Rindi S, Castellani AA (1978) Enzymatic formation of UDP-*N*-acetylgalactosamine in epiphysial-plate cartilage. Connect Tissue Res 6: 83–88
20. De Luca G, Rindi S, Castellani AA (1978) Biosynthesis of glycosaminoglycan precursors: the role of UDP-*N*-acetylglucosamine 4'-epimerase in bovine cornea. Bull Mol Biol Med 3: 260–270
21. Castellani AA, De Luca G, Rindi S, Salvini R, Tira ME (1986) Regulatory mechanisms of UDP-glucuronic acid biosynthesis in cultured human skin fibroblasts. Ital J Biochem 35: 296–303
22. Aureli G, Rizzotti M, Balduini C, Castellani AA (1969) Variazioni di mucopolisaccaridi durante il processo di invecchiamento della cartilagine. Inibizione della uridin difosfoglucoso deidrogenasi con UDP-xiloso. Riv Istochem Norm Pat 15: 9–25
23. Wolf B, Gressner AM, Nevo Z, Greiling H (1982) Age-related decrease in the activity of UDP-xylose: core protein xylosyltransferase in red costal cartilage. Mech Aging Dev 19: 181–190
24. Scott JE (1989) The chemical morphology of keratan sulphate proteoglycans. In: Greiling H, Scott JE (eds) Keratan sulphate. Chemistry, biology, chemical pathology. The Biochemical Society, London, pp 122–134
25. Bettelheim F, Goetz D (1976) Distribution of hexosamines in bovine cornea. Invest Ophtalmol 15: 301–304
26. Scott JE, Haigh M (1988) Keratan sulphate and the ultrastructure of cornea and cartilage: a "stand ind" for chondroitin sulphate in conditions of oxygen lack? J Anat 158: 95–108
27. Fatt I (1978) Physiology of the eye. Butterworth, Boston, pp 92
28. Scott JE, Stockwell RA, Balduini C, De Luca G (1989) Keratan sulphate: a functional substitute for chondroitin sulphate in O_2 deficient tissues? In: Greiling H, Scott JE (eds) Keratan sulphate. Chemistry, biology, chemical pathology. The Biochemical Society, London, pp 88–93
29. Hook M, Lindhal U, Backstrom G, Malmstrom A, Fransson LA (1974) Biosynthesis of heparin: III. Formation of iduronic acid residues. J Biol Chem 249: 3908–3915
30. Malmstrom A, Fransson LA, Hook M, Lindhal U (1975) Biosynthesis of dermatan sulphate: I. Formation of L-iduronic acid residues. J Biol Chem 250: 3419–3425

Arteriosclerosis and Aging. Distribution Pattern of Atherosclerosis in Various Arteries*

J. Metz

Introduction

Considering that parallel to a progressive decrease of premature death attributable to, e.g., infections, greater survival into very old age unmasks age-related alterations of tissues, then arteriosclerosis seems to be an aging phenomenon [4, 11]. Since we are the only species which suffers from arteriosclerosis, it appears that this disease is a consequence of a generalized evolutionary development related to aging.

Arteriosclerosis ensues when the normal regulation of structure and function of the arterial wall is disrupted. This process is seen as a final common pathway of forces both genetic and acquired that threaten the integrity of the arterial intima and influence the progression of plaques. Over the past 30 years, however, it has been dramatically proven that arteriosclerotic lesions in various vascular regions such as in the aorta, coronary or carotid arteries are clinically relevant not just in the aged human. Multiple "risk factors" have been considered to accelerate the time course of arteriosclerosis and to influence its predominate localization [9, 25]. Initiating forces, apart from inherent or age-related phenomena, can generally be divided into two categories: chemical, e. g., hypercholesterolemia, and mechanical, e. g., hypertension. Because of complex interactions leading to variable cellular compositions of the arterial wall lesions, several theories about their pathogenesis have been proposed, for example: reaction to injury [21, 23], monoclonal [3], lysosomal [31], and clonal senescence [14] theories. All provide a potential consonance, since they are compatible with both a time-related process and theories of primary aging.

Epidemiologic, clinical, and cellular research suggests a central role for lipoproteins in the process of atherosclerosis in man [9]. Thus, when the intimal barrier is injured by exposure to high levels of low-density lipoproteins (LDL), its repair mechanisms become rate limiting. Since the residual capacity for cellular replication declines in all cells as a function of age, the efficiency of repair processes within the vessel wall may progressively diminish with advancing age.

---

*Supported by the Deutsche Forschungsgemeinschaft (DFG-SFB 320) and Forschungsschwerpunkt des Landes B. W. (Altersforschung).

Reihe der Villa Vigoni, Band 1
Biology of Aging
Zwilling and Balduini (Eds.)

In animal models such as the cholesterol-fed rabbit [1], atheromas can be induced experimentally. Although differences to atherosclerosis in man have been reported, similarities of cellular processes within the plaques are obvious [27, 29].

At least four cell types are involved in the genesis and progression of atherosclerotic plaques [22, 23]. Two are endothelial cells and smooth muscle cells of the arterial wall whose homeostasis is altered, e. g., due to interactions with elevated levels of cholesterol and lipoproteins. The monocyte-macrophage system and platelets become involved by processes such as activated macrophages invading into the intima and changing into foam cells or activated platelets contributing to thrombotic events on the surface of plaques [6, 13, 19, 22]. Both latter systems might also interact via mediators such as eicosanoids (thromboxane, TXA_2) or via growth factors (platelet and macrophage derived growth factors) with endothelial and smooth muscle cells [17].

Since the hypercholesterolemic rabbit is considered to be especially useful in the study of the pathophysiology of atherosclerosis in greater detail as well as the effects of drugs on its progression, we investigated the pattern of atherosclerosis in its aorta, coronary and carotic arteries with a new computerized morphometric method.

Materials and Methods

A total of 20 male WNZ rabbits (Savo, Medizinische Versuchstierzuchten, Kisslegg, FRG) with a body weight of 1.5 – 2.0 kg were used in our experiments. Ten animals were fed a standard chow throughout the experiment; the remainder received a 0.5 % cholesterol-enriched diet (Ch-D; Altromin C 2 000, Altromin GmbH, Lage, FRG) for 96 days.

Tissue Preparation

The rabbits were anesthetized with 2 ml Narcorene (320 mg pentobarbital sodium, I.F.F.A., Merrieux GmbH, Lampheran) and exsanguinated via the axillary vessels after 96 days. The aorta, right and left common carotid arteries (CCA), and heart were removed and rinsed in 0.9 % NaCl solution. The aorta was divided into 10 equally sized segments [20], while the CCA were each divided into 25 equally sized segments. In the hearts, the left coronary artery (LCA) and the anterior interventricular branch (LAD) were dissected. The LAD was further divided in a proximal-distal orientation into 6 equally sized segments.

The specimens were fixed in 4 % buffered formalin for 3 h (aorta and CCA) or 12 h (heart). The specimens were dehydrated in 70 %, 80 %, 90 %, and 96 % ethanol, infiltrated, and embedded in an oriented manner to achieve 5-μm cross-sections in glycol-methacrylate (Historesin, LKB) which were stained with H&E and analyzed under the light microscope (Olympus BH2). A computerized image analysis system (Logos 386 computer and Matrox MVPAT board) was used for morphometric evaluation [16]. Cross-sections of the investigated vessels were magnified 8- to 40-fold, recorded via a video camera (Sony, DXC102), digitalized, and evaluated using a newly developed software program (Pavlov, unpublished thesis). The following parameters were measured: length of the intima, plaque extension (mm), and protrusional area (mm^2) of plaques. Using these values the vessel wall surface covered by plaque and luminal stenosis were calculated [20].

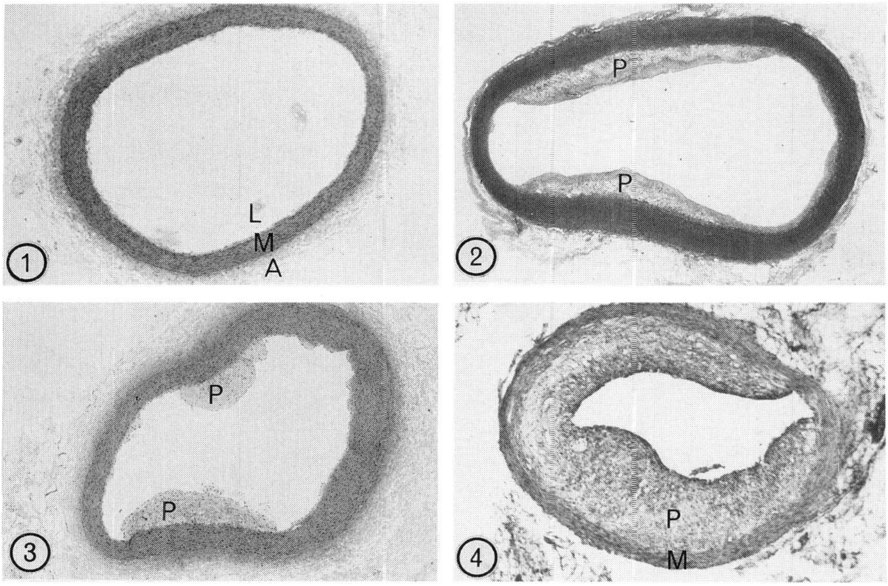

Fig. 1. Proximal thoracic aorta of an animal fed on a standard diet. Lumen (*L*) is bordered by endothelial cells; endothelium as well as intima are so thin that they cannot clearly be distinguished at this magnification. Media (*M*) consists mainly of smooth muscle cells and elastic fibers. Adventitia (*A*), which is mostly removed due to dissection of the vessel, surrounds the outside of the media (× 8)

Fig. 2. Aortic arch of rabbit fed on a 0.5 % cholesterol-enriched diet for 96 days. Significant hyperplasia (*P*) of the intima. Note the varying cellular composition within the plaque (× 5)

Fig. 3. Proximal common carotid artery of rabbit fed on a 0.5 % cholesterol-enriched diet for 96 days. Several plaques (*P*) protrude into the lumen (× 25)

Fig. 4. Proximal left coronary artery of rabbit fed on 0.5 % cholesterol-enriched diet for 96 days. Huge plaque (*P*) gives rise to a significant luminal stenosis. *M* media (× 25)

Results

Histology of Atherosclerotic Lesions

Rabbits fed standard chow (A) did not acquire any atherosclerotic lesions within the aorta (Fig. 1). After 96 days on Ch-D, all animals exhibited atherosclerotic plaques at least within the proximal regions of the vessels investigated (Figs. 2-4). The extent of atherosclerosis varied considerably among animals and within vessels. In the proximal regions of the aorta, coronary arteries, and CCA, the entire circumference of the intima often exhibited hyperplasia. The atheromatous tissue interposed between the endothelium and unaltered smooth muscle cells of the media showed a quite variable cellular constitution according to the size of the plaques and their location within the vessels. In the aorta, layers of more or less extended clusters of lipid-laden cells (foam cells) and considerable amounts of fibrillar ground substance were evident within large plaques (Fig. 5). Occasionally, cholesterol crystals, necrotic as well as calcified foci could be seen there. In smaller plaques, foam cells were mostly intermingled in the ground substance within the plaques. Emerging plaques were often seen as fatty streaks consisting mainly of foam cells (Fig. 6).

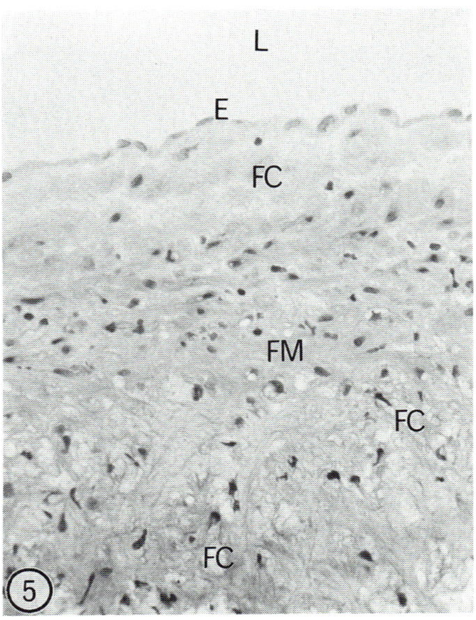

Fig. 5. Luminal two-thirds of a plaque from the proximal thoracic aorta of a rabbit fed on a 0.5 % cholesterol-enriched diet for 96 days. Complete endothelial cell layer (*E*) lines the lumen (*L*). Layers of foam cells (*FC*) are located subendothelially and clusters of them are found deeper within the plaque. *FM* fibrillar material ($\times$ 300)

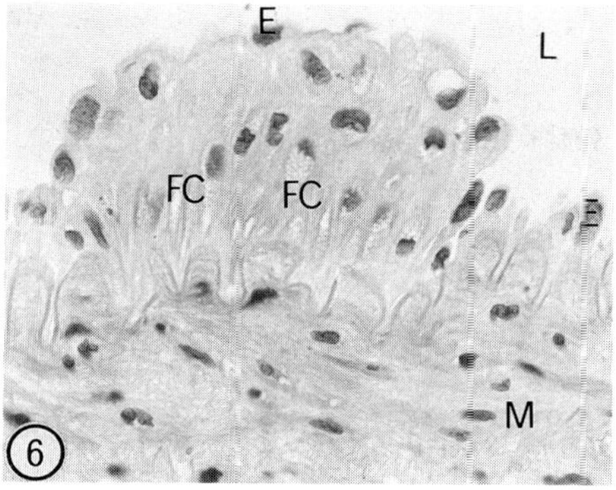

Fig. 6. Distal thoracic aorta of a rabbit fed on a 0.5 % cholesterol-enriched diet for 96 days. A fatty streak is seen containing numerous multivesiculated foam cells (*FC*). *L*, lumen; *E*, endothelial cells; *M*, smooth muscle cells in the media (× 500)

Morphometry of Atherosclerotic Lesions

A regional pattern of atherosclerosis with a proximal-distal decrease of surface covered with plaque and luminal stenosis was obvious in the aorta, LCA, and CCA (Fig. 7; Tables 1, 2). In the coronary system, the LCA always exhibited plaques leading to significant luminal stenosis (Table 1). The LAD, however, showed a quite variable proximal-distal pattern. In the CCA, the extent of atherosclerosis was more consistent throughout the five investigated segments in comparison with the LAD (Table 2).

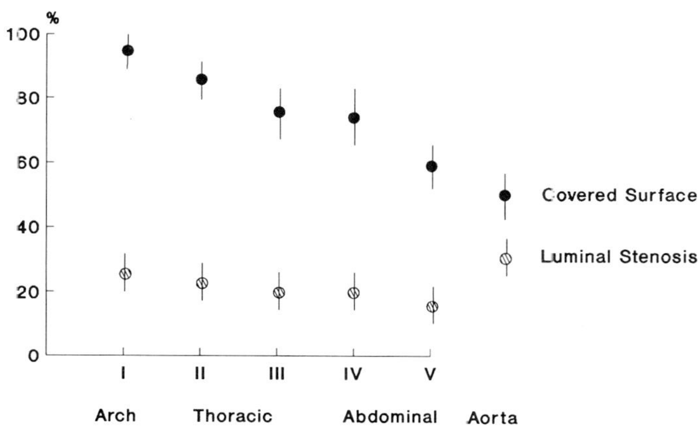

Fig. 7. Covered surface (*solid circles*) and luminal stenosis (*hatched circles*) of aortic regions (mean values ± SEM) in WNZ rabbits ($n = 10$) fed 0.5 % cholesterol-enriched diet for 96 days

Table 1. Atherosclerosis in coronary vessels of WNZ rabbits fed 0.5 % cholesterol-enriched diet

Vessel	Animals (*n*)	Frequency of plaques (%)	Surface covered by plaque (%)	Luminal stenosis (%)
Left coronary artery (LCA)	7	100.0	82.0	46.7
Anterior interventricular branch (LAD)				
Segment 1	9	44.0	13.4	6.6
Segment 2	8	12.5	2.2	0.3
Segment 3	9	44.0	14.9	3.4
Segment 4	7	14.0	4.5	1.1
Segment 5	8	12.5	5.4	1.9
Segment 6	6	16.0	9.8	2.6

Table 2. Atherosclerosis in common carotic arteries of WNZ rabbits fed 0.5 % cholesterol-enriched diet

	Animals (*n*)	Frequency of plaques (%)	Surface covered by plaque (%)	Luminal stenosis (%)
Segment 1	9	88	33.2	4.6
Segment 2	8	78	20.9	11.0
Segment 3	9	70	16.0	5.8
Segment 4	7	80	14.4	12.6
Segment 5	8	88	14.3	13.0

Discussion

The cholesterol-fed rabbit is one of the most-used experimental models for hypercholesterolemia and atherosclerosis [1]. Employing this model, highly variable extents of atheromas within the aorta, coronary and carotic arteries of individual animals were found, although they were kept under identical conditions. Despite this variability, it was apparent in all animals that the amount of plaques decreased in each vascular region in a proximal-distal pattern. The decrease, however, was not linear, e. g., from the aortic arch to the abdominal aorta [16]. This may indicate that the distribution pattern of plaques is influenced by other factors: Lesion-susceptible areas as well as raised shear forces and turbulence of the bloodstream injuring the vessel wall at divisions to a greater extent have been proposed as candidates [5].

The general cellular constitution of the atheromas was similar in all Ch-D-fed animals, exhibiting a considerable variation of cellular components according to their size, which to a certain extent reflected their localization within the vessels. Since different constitutions and cellular compositions of the induced atheromas were reported to depend on the amount of cholesterol in the diet [29, 30], the hypercholesterolemic rabbits in our study exhibited arterial wall lesions characteristic for a rather rapid progression of atherosclerosis. They showed features of lipid atheromas containing numerous "foam cells" typical of an increase of cholesterol content [18].

The cellular processes which occur during atherogenesis in the cholesterol-fed rabbit include migration and proliferation of smooth muscle cells and macrophages, accumulation of foam cells, synthesis and secretion of acid mucopolysaccharides and other connective tissue components into the atheromatous plaque [10, 23]. Some of these processes are similar to an inflammatory-like response or can be subsumed under general mesenchymal reactions [10]. In the complex reactions of the vessel wall, several processes are activated by cyclooxygenase and lipoxygenase metabolites of the arachidonic acid cascade [2]. Platelets, leukocytes, and macrophages are major sources of TXA_2 [8, 12, 26], leading to its increase within the serum and within atheromas [15], suggesting a major role of this mediator in the pathogenesis of the atherosclerotic process. Furthermore, platelets, macrophages, and endothelial cells are all sources of growth factors, which are thought to induce proliferative responses, e. g., of smooth muscle cells [3].

Conclusion

Arteriosclerosis has been recognized in man as a complex disease of the arterial wall which depends on persisting risk factors. It mainly develops as an age-related process and leads to premature death or severe complications such as ischemia or infarction in heart, brain, and legs, not only in the elderly. Focal intimal lesions of inception are observed even in infants; their progression into clinically significant plaques is, however, still not clear.

In conclusion, accelerated arteriosclerosis in man as well as experimentally induced atherosclerosis, e. g., in the hypercholesterolemic rabbits, are dependent on initiating forces which finally lead to a time-related process resulting in a progressive narrowing of the arteries. In addition to acquired phenomena, in many individuals the progression of arteriosclerosis might be influenced by age-related processes of the vessel wall due to the inherent regulation of gene expression. Intervention by drugs has the aim to slow down the progression of accelerated arteriosclerosis (e. g., [2, 18, 24, 29]) or prevent complications [7].

Acknowledgements: The technical help of I. Stenull, R. Hertel, and Ch. Willich in laboratory work is acknowledged.

References

1. Anitschkow N, Chalatow S (1913) Über die experimentelle Cholesteatose und ihre Bedeutung für die Entstehung einiger pathologischer Prozesse. ZB Allg Path 24: 1 – 9
2. Bailey JM, Butler J (1985) Anti-inflammatory drugs in experimental atherosclerosis. Part 6: combination therapy with steroid and non-steroid agents. Atherosclerosis 54: 205 – 12
3. Benditt P, Benditt JM (1973) Evidence of a monoclonal origin of human atherosclerotic plaques. Proc Natl Acad Sci USA 70: 1753
4. Doerr W (1987) Die Pathologie Virchows und die Lehre von der Arteriosklerose. Pathologie 8: 1 – 8
5. Faggiotto A, Ross R, Harker L (1984) Studies of hypercholesterolemia in the nonhuman primate: I. Changes that lead to fatty streak formation. Arteriosclerosis 4: 323 – 340
6. Fowler SD, Mayer EP, Greenspan P (1986) Foam cells and atherogenesis. Ann NY Acad Sci 79 – 90
7. Gaetano G (1988) Primary prevention of vascular disease by aspirin. Lancet 14: 1093 – 1095
8. Goldstein IM, Malmsten CL, Kindahl H, Kaplan HB, Rademark O, Samuelsson B, Weissman G (1978) Thromboxane generation by human peripheral blood polymorphonuclear leukocytes. J Exp Med 148: 787 – 792
9. Gordon R, Kannel WB (1972) Predisposition to atherosclerosis in the heart and legs. The Framingham Study. JAMA 221: 661 – 666
10. Hauss WH (1979) The role of arterial wall cells in atherogenesis. Cardiovasc Res 17: 75
11. Hazzard W (1985) Aging and atherosclerosis. Clin Geriatr Med 1: 251 – 284
12. Higgs GA, Bunting S, Moncada S, Vane JR (1976) Polymorphonuclear leukocytes produce thromboxane A_2-like activity during phagocytosis. Prostaglandins 12: 749 – 757
13. Joris I, Zand T, Nunnari JJ, Krolikowski FJ, Majno G (1983) Studies on the pathogenesis of atherosclerosis: I. Adhesion and emigration of mononuclear cells in the aorta of hypercholesterolemic rats. Am J Pathol 113: 341 – 358
14. Martin G, Sprague C (1973) Symposium on in-vitro studies related to atherogenesis: life histories of hyperplastoid cell lines from aorta and skin. Exp Mol Pathol 18: 125 – 141
15. Mehta JL, Lawson D, Mehta P, Saldeen T (1988) Increased prostacyclin and thromboxane A_2 biosynthesis in atherosclerosis. Proc Natl Acad Sci USA 85: 4511 – 4515
16. Metz J, Schmelz A, Pavlov P (1990) Morphologische Untersuchungen zur Quantifizierung der Arteriosklerose. Z Gerontol 23: 140 – 142
17. Mullane KM, Bradley G, Moncada S (1982) The interactions of platelet-derived mediators on isolated coronary arteries. Eur J Pharmacol 84: 115 – 118
18. Pill J, Wolf O, Schmelz A, Stegmeier KH, Metz J (1990) Investigations of the antiatherosclerotic effect of daltroban. Z Kardiologie 79 [Suppl. 3]: 155 – 160
19. Pittilo RM, Machin JS (1988) Platelet-vessel wall interactions. Springer, Berlin Heidelberg New York, pp 1 – 256
20. Polimeni PJ, Cunningham P, Otten MD, McCrea D (1987) Morphometric quantification of atherosclerotic plaques by computer-assisted image-analysis of histographs. Comp Biomed Res 20: 113 – 124
21. Ross R (1981) Atherosclerosis: a problem of the biology of arterial wall cells and their interactions with blood components. Arteriosclerosis: 1: 293 – 311
22. Ross R (1986) Endothelium, monocytes, platelets, and atherosclerosis. Monogr Atheroscler 14: 169 – 172
23. Ross R (1986) The pathogenesis of atherosclerosis – an update. N Engl J Med 314: 488 – 500
24. Saunders RN (1982) Evaluation of platelet-inhibiting drugs in models of atherosclerosis. Annu Rev Pharmacol Toxicol 22: 279 – 95
25. Schettler G (1962) Atherosklerose: Ätiologie, Pathologie, Klinik und Therapie. Thieme, Stuttgart

26. Siess W, Cuatrecasas P, Lapetina EG (1983) A role for cyclooxygenase products in the formation of phosphatidic acid in stimulated human platelets. Differential mechanisms of action of thrombin and collagen. J Biol Chem 258: 4683 – 4686
27. Small DM (1988) Progression and regression of atherosclerotic lesions. Arteriosclerosis 8: 103 – 129
28. Stehbens WE (1986) An appraisal of cholesterol feeding in experimental atherogenesis. Prog Cardiovasc Dis 29: 107 – 128
29. Velican C (1974) Macromolecular changes in atherosclerosis. In: Graumann W, Wenmann K (eds) Handbuch der Histochemie. Fischer, Stuttgart, pp 5 – 21
30. Vesselinovitch D (1979) Animal models of atherosclerosis, their contributions and pitfalls. Artery 2: 193 – 206
31. Wolinsky H, Fowler S (1978) Participation of lysosomes in atherosclerosis. N Engl J Med 299: 1173 – 1178

Physiological and Surgical Aspects of the Geriatric Patient

R. Dionigi and A. Benevento

Introduction

Aging is a normal and universal process; therefore, it is not a pathological condition but does predispose to many diseases. Over the years, we observed on the one hand a progressive decrease in birth indices and on the other an increase in life expectancy. The average length of life in Italy has increased to 78 years for women and to 71 years for men. In the world, 10 % of the population is over 65 years old.

As a direct consequence of the increase in life expectancy, the most frequent diseases leading to death have changed, being chronic and degenerative diseases instead of acute diseases. The most common causes of death are, in fact, cardiovascular and neoplastic disease [1].

In a healthy person over 65 years old, the basal metabolism rate is reduced to 40 % compared with a 30-year-old person (Fig. 1). This is due to a progres-

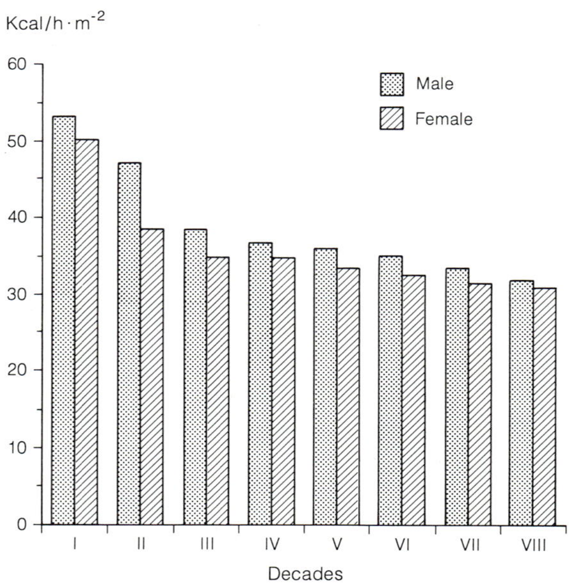

Fig. 1. Basal metabolic rate at different ages

Reihe der Villa Vigoni, Band 1
Biology of Aging
Zwilling and Balduini (Eds.)

sive age-related reduction in the functioning of many organs, major losses including renal, liver, pulmonary, and immune activities. Hence, in all aging organisms there is a steadily declining ability to adapt to changes in the surrounding environment and stressful events such as surgery.

The high surgical risk of these patients depends on age itself, their general condition, of the disease, presence of concomitant diseases, and risks related to anesthesiological drug administration. Considering this high surgical risk, many efforts have been made to perform proper and more aggressive surgery in a short time, thanks to ever more sophisticated techniques. Major improvements have been achieved in cardiovascular and oncologic surgery.

Physiological Aspects

The elderly are not simply adults with a greater chance of having coexisting disease; they are fundamentally different. Their general metabolism, cardiovascular and respiratory physiology, functioning of the liver and kidney, and the immune response are deeply modified. Other minor impairments involve thermoregulation, muscle strength and loss of tissue with a proportional increase in fatty tissue. This last aspect may contribute to altering drug distribution together with the impaired liver and renal function. Moreover, we observed a decrease in pancreatic and small intestine function. Older people are found to have a significant decrease in bicarbonate, lipase, and amylase output. The alteration of the adaptative response of intestinal and pancreatic function to starvation and refeeding may have an adverse impact upon nutritional maintenance [2].

Cardiovascular System

Left ventricular hypertrophy is an adaptative response to the increase in cardiac overload caused by age-related arterial dilatation and loss of vascular compliance. In the large arteries, clinically significant intimal thickening and loss of distensibility occur, resulting in increased pulse pressure through elevation of systolic pressure [3, 4].

Systolic hypertension may also contribute to the atherosclerotic process. With aging, progressive dilatation and elongation of the aorta occurs, accompanied by a gradual stiffening of the aortic wall and reduction in its compliance, due to elastic fiber degeneration from either collagen and muscular bundle loss.

According to recent investigations the degenerative changes of the arterial wall tend to stabilize in subjects over 60 years of age in all arteries except those of the cerebral region, where the lesions tend to show a continuous progression related to age [5]. Immunohistochemical studies demonstrated the predominance of α-adrenoreceptor-mediated effects on many arteries, with a subsequent vasoconstriction [3].

Respiratory System

Respiratory function in the elderly is impaired as noted by reduced muscular contraction and lower vital capacity. On the contrary the functional residual capacity is increased. The ventilatory response to hypercarbia and hypoxia at the age of 70 years is half that of young persons, and laryngeal, pharyngeal, and airway reflexes are all less effective [6].

Most studies of blood gas concentration have shown a linear decrease in oxygen tension with age. This would suggest, even in the absence of disease in advanced age, that the subjects are hypoximic [7].

Liver Function

It has been shown that age is correlated with a negative effect upon liver volume and blood flow to the liver [8]. Tables 1 and 2 show the percentage of decline from young to older subjects. These alterations suggest that liver function impairment is mostly due to the blood component than to loss of parenchymal elements. However, as well as the marked changes in perfusion, decreased hepatic enzyme activity has been observed with an impairment of certain metabolic pathways.

The reduction in liver volume and perfusion may at least partly account for the decline in the clearance of many drugs undergoing liver metabolism. The lower liver function is also responsible for the decreased serum albumin levels observed in elderly people.

Table 1. Liver volume changes (ml) with age

	24-year-old subject	91-year-old subject	Fall (%)
Males	1524	1102	28
Females	1415	789	44
Mean	1474	934	37

Table 2. Apparent liver blood flow changes with age

	24-year-old subject (ml/min)	91-year-old subject (ml/min)	Fall (%)
Males	1864	1126	40
Females	1564	645	50
Mean	1717	807	53

Table 3. Urinary volume rate changes with age

	Volume rate (ml/day)
Newborn	17
6 – 14 years	444
Adults	
Males	1360
Females	1130
>90 years	853

Table 4. Immunoglobulin concentration changes with age

	IgG (mg/dl)	IgM (mg/dl)	IgA (mg/dl)
Young	1064	154	166
Elderly	1149	106	203
Normal range	640 – 1420	50 – 280	60 – 340

Renal Function

Renal blood flow and glomerular filtration rate decrease by about 1% per year; renal acidification and tubular secretion rate are both lowered in the elderly [9]. The urinary volume rate (ml/day) decreases greatly over the age of 90 years (Table 3). The rate of creatinine clearance declines with aging and represents a useful guide to the rate of renal drug elimination in the elderly.

Immune Response

Older subjects can be considered immunodepressed for humoral and cellular responses. Serum concentration of IgM shows a statistically significant fall in the elderly (Table 4) [10].

Immune system alterations with age include a progressive decline in antibody response against exogenous antigens and a gradual increase of incidence of autoimmune events (Fig. 2) [11]. These changes seem to play an important role in the biological aging process.

T lymphocytes undergo a clear functional impairment. Moreover, a decrease of T helper/T suppressor rate has been observed, mainly due to a reduction of the T-helper subpopulation.

Older subjects are consequently greatly predisposed to infectious diseases. Infections represent a significant threat in surgical patients. The probability of developing a postsurgical infection is very high in the elderly, particularly if

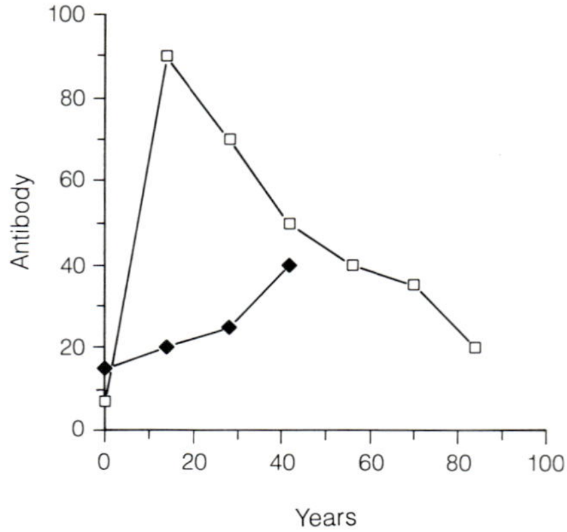

Fig. 2. Effect of aging on antibody reactivity (□ , natural exogenous antigen antibodies in human serum; ◆ , antinuclear antibody level)

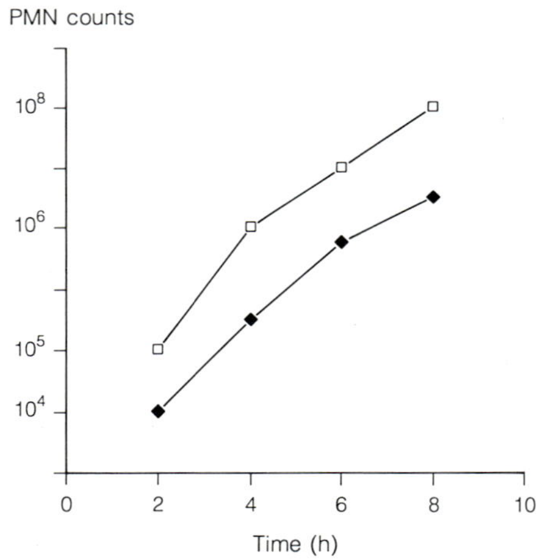

Fig. 3. Neutrophil delivery into inflammatory skin (□ , young; ◆ , elderly)

associated with malnutrition. Malnutrition negatively influences neutrophil (PMN) function. The delivery of PMN into inflammatory skin is significantly reduced in the elderly (Fig. 3). In contrast the concentration of complement factors C3 and C4 is significantly higher in older people (Table 5) [10].

Based on these physiopathological considerations, many surgical techniques have been used in order to avoid this threatening complication. Of particular importance are preparation of the patient before surgery, strict observation of aseptic rules, antibiotic prophylaxis, and last but not least nutritional support.

Table 5. Complement concentration changes with age

	CH-50	C3 (mg/dl)	C4 (mg/dl)
Young	168	133	26.7
Elderly	214	158	41.9
Normal range	110 – 190	100 – 233	15 – 48

$P = 0.04 - 0.005$ for all measurements

Surgical Aspects

Geriatric patients represent an important class in our society. The elderly require more surgical services than ever before and expect high quality surgical and medical care.

In Italy, the proportion of surgical operations performed on elderly patients rose from 28 % of the general population in 1980 to 42 % in 1989. In our series we observed an overall peak of surgical procedures in the 7th decade (30.82 %) (Fig. 4).

The most common diseases needing surgical procedures in patients over 65 years old were in order of decreasing frequency: vascular disease (aortic aneurysm, peripheral vascular disease, carotid stenosis, deep venous thrombosis, acute inferior limb ischemia) 11.09 %, gallstone 10.72 %, hernias 10.18 %, gastric cancer 8.18 %, and colorectal cancer 7.27 %.

Important advances have been achieved in vascular surgery, with particular regard to carotid occlusive disease; for example, it has become possible to perform thromboendoarterectomy under local anesthesia, thus allowing continuous check of the patients active response and circumventing the potential onset of cerebral ischemic complications and anesthesiological risks.

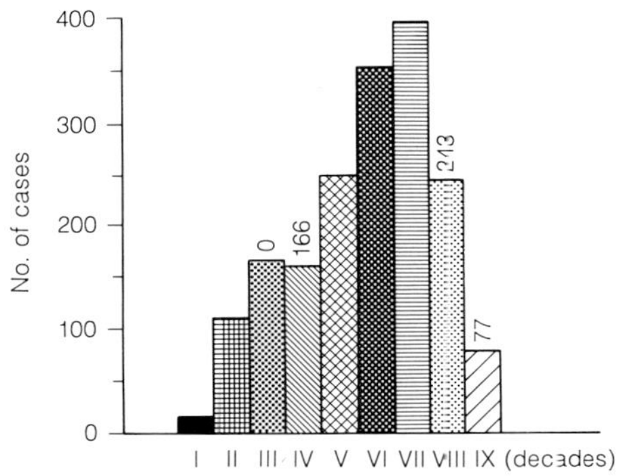

Fig. 4. Number of surgical procedures in various age groups

An other improvement in vascular surgery involves new graft materials, especially in the treatment of abdominal aortic aneurysms, which allow very safe stitching procedures.

In addition to conventional operations in high-risk patients, extraanatomical surgery has been proposed (femoro-femoral bypass, axillo-femoral bypass), in order to realize lesser traumatic and quicker operations.

Oncologic surgery has also gained in importance. Many reasons explain the high incidence of geriatric patients affected by neoplastic disease who undergo surgical resection: longer life expectancy, longer exposure to carcinogenic agents, immune system impairment, and more aggressive radical and pallative surgery, with the help of revolutionary and sophisticated instruments.

Major improvements have been achieved in esophageal and rectal surgery with the employment of mechanical stapler techniques. These devices allow the surgeon to resect under easier conditions, especially when tumor sites are very deep and narrow for anatomical reasons. In this way, it is possible to perform radical operations, resecting the tumor with safe resection margins. Moreover the procedure duration is greatly shortened, with many advantages for postoperative recovery: shorter exposure to anesthesia, lower dose administration, earlier mobilization, and reduced incidence of infectious and non-infectious complications.

With the advent of mechanical staplers it has been possible to perform major surgical operations on the pancreas even in elderly patients. Resection of the pancreatic proximal portion by mechanical staplers permits a shorter resection time and lowers the risk of severe postoperative complications. Moreover, the presence of the residual distal portion preserves a good pancreatic endocrine function.

An new intraoperative staging technique, first tried 5 years ago in the USA, has been recently introduced into Italy. It is called radioimmunoguided surgery and consists of the preoperative administration of radiolabelled monoclonal antibodies which selectively bind to a specific tumor-associated antigen. At the time of surgery, by means of a hand-held gamma-detecting probe, it is possible to scan the entire abdomen, looking for radioactive sources. The specificity and resolution of this technique is extremely high, being able to identify tumor deposits only 1 mm in diameter. The surgeon can outline tumoral resection margins and secondary localizations and therefore adjust the surgical procedure according to the data obtained from the probe itself. In our experience, it has been possible to modify the preoperatively planned surgical procedure during the operation in 15 % of cases.

If on the one hand it is very important to perform high-quality surgery, exploiting every advanced device, the incidence of postoperative complications still remains high in this group of patients. The most threatening complication is the onset of infectious disease in the postoperative period. It has been estimated that the incidence of infectious complication is higher if associated with malnutrition. As malnutrition is more frequent in elderly patients, this further helps to increase the risk. We correlated the incidence of surgical infection with age (Table 6).

Table 6. Incidence of surgical infections by age

	Patient's age (years)		
	14 – 35	36 – 65	>65
Infections (n)	45	220	202
Total no. of patients (n)	356	932	508
Infection incidence (%)	12.6	23.6	39.8
	$P=0.00041$		$P=0.0000038$
		$P=0.000001$	

Table 7. Incidence of postoperative surgical infections by age

	Patient's age (years)		
	14 – 35	36 – 65	>65
Wound infection	9	29	36
Respiratory tract infection	3	22	23
Urinary tract infection	0	14	27
Intraabdominal abscess	0	2	3
Peritonitis	0	0	0
Others	0	0	3

In our experience a statistically significant increase of incidence of infections with age was demonstrable. This complication occurred in 12.6 % of patients between 14 and 35 years old, in 23.6 % of those between 36 and 65 years old, and in 39.8 % of those over 65 years old (χ^2 test: $P=0.000001$).

The most common infectious complications involved: the wound itself, respiratory tract, urinary tract, intraabdominal abscess (Table 7). Among these, wound infection is the most frequent. In "clean" surgical operations the percentage of elderly patients with onset of wound infection in the postoperative recovery phase was double that in younger ones. The proportion triples in "contaminated" procedures (Table 8). As mentioned above, the reasons why elderly patients are predisposed to infectious disease are to be found in their immune system impairment. The final result is that the wound healing process is prolonged. The physiologic decline of aged skin involves mechanical, biochemical, and trophic changes [13]. The water content of connective tissue is reduced, microcirculation and tissutal O_2 are decreased, and degenerative changes weaken the elastic fiber, facillitating the formation of tortuosities around collagen bundles, with a resulting loss of elasticity and hydration of the skin and dermis.

Table 8. Percentage of wound infections by age

Classification of surgical infection[a]	Patient's age (years)		
	14 – 35	36 – 65	>65
Clean	1.6 %	1.8 %	3.6 %
Clean-contaminated[b]	0.8 %	3.5 %	10.3 %
Contaminated	8.1 %	12.8 %	21.1 %
Dirty	7.9 %	8.0 %	15.4 %

[a] From [12]

[b] Clean-contaminated are those surgical procedures where bileary, urinary, or digestive tract are opened, but with minimal possible contamination.

In conclusion, an ever increasing number of surgical patients is made up of elderly people. With aging, most organs and systems show a functional decline and a consequent reduction of the physiological fitness. This aspect represents an important risk factor for surgery. Many efforts have been carried out in order to improve surgical procedures and the clinical outcome of these patients, who easily develop severe complications. Major advances have been achieved in vascular and oncologic surgery, which is potentially able to offer very high surgical standards.

References

1. Kennedy BJ (1988) Aging and cancer. J Clin Onc 6 (12): 1903 – 11
2. Wellas BJ, Belas D et al. (1990) Adaptative response of pancreatic and intestinal function to nutritional intake in the aged. JAGS 38 (3): 254 – 258
3. Safar M (1990) Ageing and its effects on cardiovascular system. M Drugs 39 (1): 1 – 8 (44 ref)
4. Kelly R, Hayward C et al. (1989) Noninvasive determination of age-related changes in the human arterial pulse. Circulation 80 (6): 1652 – 1659
5. Weber G, Bianciardi G et al. (1988) Atherosclerosis and aging. Arch Pathol Lab Med 112 (10): 1066 – 70
6. Jones MR (1989) Anaesthesia in old age. 44 (5): 377 – 378
7. Blom H, Mulder M et al. (1988) Arterial oxygen tension and saturation in hospital patients: effect of age and activity. Br Med J 297 (9): 720 – 721
8. Wynne HA, Cope LH et al. (1989) The effect of age upon liver volume and apparent liver blood flow in healthy man. Hepatology 9 (2): 297 – 301
9. Montamat SC, Cusack BJ et al. (1989) Management of drug therapy in the elderly. New Engl J Med 321 (5): 303 – 309
10. McGregor RR, Shalit M (1990) Neutrophil function in healthy elderly subjects. J Gastroent 45 (2): M55 – 60
11. Doria G, Frasca D (1989) Immunologia dell'invecchiamento. In: Dammacco F (ed) Immunologia e Medicina. Ermes, Milan, pp 239 – 252
12. Altemeier WA, Burke JF, Pruilt BA, Sandusky NR (1976) Definition and classifications of surgical infections. In: Altemeier WA (ed) Manual on control of infections in surgical patients. Lippincott, Philadelphia, p. 6
13. Imayama S, Braverman IM (1989) A hypothetical explanation for the ageing of skin. Am J Pathol 134 (5): 1019 – 25

Poly(ADP-ribosyl)ation, Genetic Instability, and Aging

A. Bürkle, K. Grube, and J.-H. Küpper

Introduction

Poly(ADP-ribosyl)ation is a posttranslational protein modification catalyzed by poly(ADP-ribose) polymerase (PARP), a highly conserved nuclear enzyme which uses nicotinamide-adenine dinucleotide (NAD) as substrate (for review, see [1]). The DNA-binding domain of the enzyme specifically binds to DNA single- or double-strand breaks, resulting in enzyme activation. Thus, treatment of cells with chemical or physical carcinogens induces a dose-dependent stimulation of polymer synthesis and turnover. To understand the biological function(s) of poly(ADP-ribosyl)ation, NAD analogues have been extensively used as competitive inhibitors of poly(ADP-ribosyl)ation in intact cells. Such inhibitors (e.g.,benzamide and derivatives such as 3-aminobenzamide) have no influence on cell growth (at concentrations of 1 m*M* or lower) nor are they mutagenic or carcinogenic. They potentiate, however, cytotoxicity and chromosomal damage induced by carcinogen treatment, e.g., alkylating agents or ionizing radiation. These and other findings led to the view that poly(ADP-ribosyl)ation plays a role in DNA repair.

Poly(ADP-ribosyl)ation and DNA Amplification

DNA amplification is one manifestation of genomic instability which plays a pivotal role in several steps of tumor development (for review see [2, 3]) as well as in aging processes of fungi (e.g. *Podospora anserina*; see chapter by H.D. Osiewacz, this volume) and mammalian cells (for review, see [4]). Since DNA amplification can be greatly induced in cell culture systems, e.g. by carcinogens, we asked the question whether poly(ADP-ribosyl)ation as an immediate response to (carcinogen-mediated) DNA strand breakage has any influence on the amplification process.

Note the abbreviations used are: *3AB*, 3-aminobenzamide; *AF*, amplification factor; *DHFR*, dihydrofolate reductase; *EF*, enhancement factor, *MNNG*, N-methyl-N'-nitro-N-nitrosoguanidine; *MTX*, methotrexate; *PARP*, poly(ADP-ribose) polymerase; *PE*, plating efficiency; *SV40*, Simian virus 40.

Reihe der Villa Vigoni, Band 1
Biology of Aging
Zwilling and Balduini (Eds.)

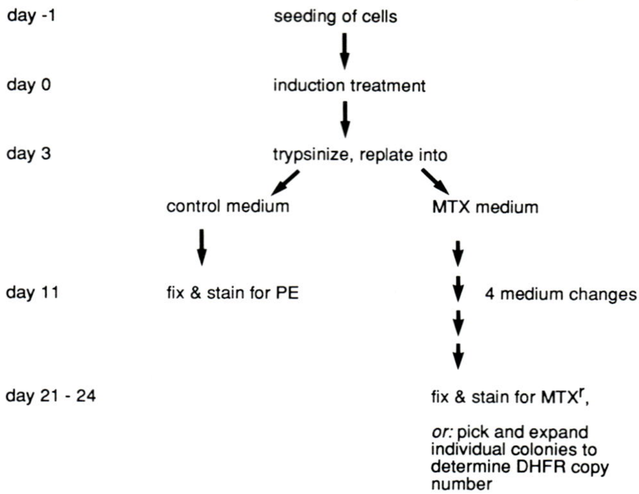

Fig. 1. Induction treatment and methotrexate (MTX) selection protocol. *PE*, plating efficiency; *DHFR*, dihydrofolate reductase

Table 1. Induction of methotrexate (MTX) resistance after treatment of Chinese hamster ovary (CHO) cells with 2 μM *N*-methyl-*N'*-nitro-*N*-nitrosoguanidine (MNNG) in the presence of increasing concentrations of 3-aminobenzamide (3AB)

3AB	MNNG	% PE	MTX^r	MTX^r/PE	EF
–	–	90.2 ± 7.0	10 ± 3.5	11	1
–	2 μM	5.1 ± 0.36	52 ± 6.6	1019	92
0.1 mM	2 μM	4.0 ± 0.33	100 ± 3.5	2500	225
0.4 mM	2 μM	3.0 ± 0.28	112 ± 24.7	3797	342
1.0 mM	2 μM	2.2 ± 0.19	123 ± 14.7	5721	515

Note: Cells were plated in 150-cm^2 culture flasks at 2×10^5 (for untreated controls) or 4×10^5 (for all MNNG treatments). The next day, cells were first given 3AB, and 1 h later MNNG was added as indicated. Three days later, cultures were trypsinized and replated for the determination of plating efficiency (PE) and MTX resistance. For PE, 500 or 2000 (control/MNNG) cells were plated in triplicate onto 10-cm Petri dishes in medium containing 10% dialyzed fetal calf serum. For the parallel determination of MTX resistance 5×10^5 cells were plated in triplicate onto 10-cm dishes in medium containing 10% dialyzed fetal calf serum and 350 nM MTX.
% PE, number of colonies × 100/number of cells plated (mean ± SD of triplicates); MTX^r, number of MTX-resistant colonies per dish (mean ± SD of triplicates); MTX^r/PE, MTX resistance frequency; EF, enhancement factor = MTX resistance frequency relative to untreated controls

We initially studied inducible DNA amplification in a simian virus (SV40)-transformed Chinese hamster cell line (CO 60) that amplifies integrated SV40 sequences after carcinogen treatment and serves as a short-term model system.

We could show that inhibition of carcinogen-stimulated poly(ADP-ribose) synthesis by 3-aminobenzamide (3AB) was correlated with a two- to six-fold potentiation of inducible DNA amplification in these cells [5, 6]. Recently, we extended those studies towards the development of methotrexate (MTX) resistance associated with dihydrofolate reductase (DHFR) gene amplification in Chinese hamster ovary cells [7]. The protocol followed is outlined in Fig. 1.

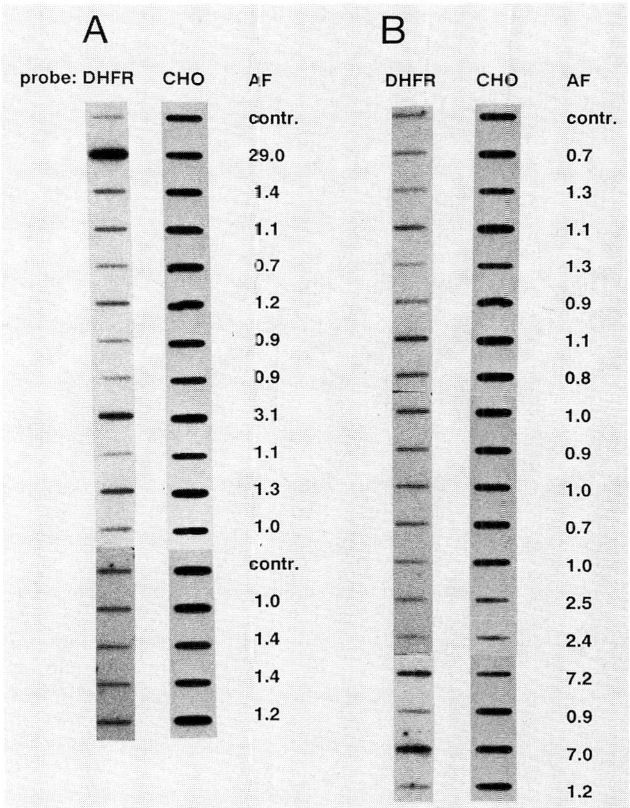

Fig. 2. Dihydrofolate reductase (*DHFR*) gene copy numbers in methotrexate (MTX)-resistant cell clones. Slot blots of DNA from individual MTX-resistant clones were hybridized with a DHFR-specific probe and subsequently rehybridized with labeled genomic CHO DNA as an internal standard, and relative copy numbers of DHFR sequences (amplification factors, *AF*) were determined. The amplification factor of controls (MTX-sensitive cells) is 1 by definition. *A**, MTX-resistant clones obtained after induction with 2 μM *N*-methyl-*N*'-nitro-*N*-nitrosoguanidine (MNNG); *B**, clones obtained after induction with 2 μM MNNG in the presence of 1 m*M* 3-aminobenzamide (3AB); *contr*, control. (For experimental details, see [7])

As is shown in Table 1, treatment with the alkylating carcinogen *N*-methyl-*N*'-nitro-*N*-nitrosoguanidine (MNNG) 3 days before selection with 350 n*M* MTX increased MTX resistance frequency by about 100-fold (enhancement factor, EF), as expected. Addition of 3AB before MNNG treatment further potentiated the frequency of MTX resistance by up to fivefold in a dose-dependent manner, parallel to a potentiation of cytotoxicity (measured as a further reduction of plating efficiency, PE). The same potentiation occurred after cotreatment with benzamide (1 m*M*), another poly(ADP-ribosyl)ation inhibitor, under conditions which precluded direct drug interactions. Benzoic acid, a noninhibitory analogue, had no effect on the MNNG-induced MTX resistance frequency. Furthermore, 3AB, benzamide, and benzoic acid, each at 1 m*M*, had no effect on the spontaneous frequency of MTX resistance. A number of individual MTX-resistant colonies were expanded to determine their DHFR gene copy number (Fig. 2). The relative frequency of DHFR gene amplification was similar (14 % vs. 22 %) whether clones were derived from cultures induced with MNNG alone (panel A) or MNNG in the presence of 1 m*M* 3AB (panel B). We thus infer that, along with the potentiation of MTX resistance frequency, 3AB potentiates the frequency of DHFR amplification to the same extent.

These and other results previously obtained with CO 60 cells [8] led us to conclude that poly(ADP-ribosyl)ation may act as a negative regulatory factor in the induction of DNA amplification (Fig. 3), since PARP inhibitors potentiated MNNG-induced SV40 DNA amplification as well as MNNG-induced MTX resistance associated with DHFR gene amplification.

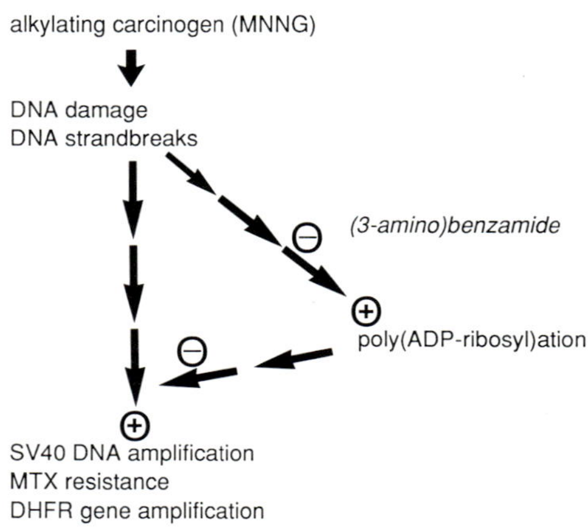

Fig. 3. Poly(ADP-ribosyl)ation as a negative regulatory factor in carcinogen-inducible DNA amplification and methotrexate (*MTX*) resistance. + denotes induction; − denotes inhibition; *MNNG*, *N*-methyl-*N*'-nitro-*N*-nitrosoguanidine; *DHFR*, dihydrofolate reductase

Poly(ADP-ribosyl)ation and Species Longevity

Several lines of investigation provide evidence that, apart from the multistep process of carcinogenesis, the aging of cells and organisms is also accompanied (if not caused) by genetic instability (for review see; [4]). In a number of biological systems, DNA excision and/or amplification events are connected with aging, e.g., in the aerobic fungus *Podospora anserina* (see chapter by H.D. Osiewacz, this volume). Furthermore, during their replicative life span, human diploid fibroblasts constantly lose certain repetitive DNA sequences dispersed over the chromosome length [9], but also telomeric DNA located at the chromosome ends [10]. Strikingly, this latter phenomenon also occurs in a yeast mutant (EST-1) which undergoes senescence, in contrast to wild-type yeast [11]. On the other hand, electron microscopic studies have revealed an increased amount of extrachromosomal circular DNA molecules in cells of aged rats and human beings [12].

Interestingly, there is a correlation between DNA repair capacity and (mammalian) life span [13–15]. DNA repair certainly constitutes a protective factor against the constant attacks by endogenous and exogenous DNA-damaging agents and thus contributes to the life-long maintenance of genome integrity and stability. In this regard, the efficient DNA repair of long-lived species could also be responsible for the delay in tumor development as compared with short-lived species. Since poly(ADP-ribosyl)ation is involved in DNA repair and the recovery from DNA damage [1], it is interesting that a highly significant positive correlation was described between the life spans of different mammalian species and PARP activity in their nucleotide-permeable mononuclear leukocytes after supralethal gamma-irradiation [16]. Physical or chemical carcinogens, however, might not induce the same number of DNA breaks if applied to living cells of different organisms, since many of the breaks are mediated by free radical mechanisms and/or DNA repair endonucleases whose activities are already known to correlate with the species' life span [13–15]. Therefore, it is not clear whether the correlation between life span and PARP activity is direct, i.e., due to a higher enzyme content or a greater specific activity, or indirect, i.e., due to other cellular functions.

In order to resolve this question and to provide a direct stimulus for PARP in permeabilized cells, we adopted the finding of Berger and Petzold [17] who showed, with activity assays of purified PARP, that double-stranded deoxyoligonucleotides are even better enzyme activators than nicked calf-thymus DNA, obviously by providing a large number of DNA ends. We recently could demonstrate that the addition of a palindromic (and hence double-stranded) decameric deoxyoligonucleotide (CGGAATTCCG) to the reaction buffer stimulated PARP activity in permeabilized Molt-3 human lymphoma cells up to about 30-fold [18]. Maximal enzyme activities were in the same range as those obtained when supralethal doses of gamma-irradiation were applied to living cells on ice immediately before the assay. Thus, we conclude that double-stranded oligonucleotides may be conveniently used as chemically and stoichiometrically well-defined PARP activators in permeabilized or ethanol-fixed mammalian cells.

Using such a modified assay, we can thus exclude any influence by other repair activities and are now retesting PARP activity as a function of the 'species' life span. Our initial results indeed indicate a significant positive correlation between the life span of 11 mammalian species and the maximal PARP activity in their peripheral blood mononuclear cells (K. Grube and A. Bürkle, unpublished observations).

Overexpression of the Poly(ADP-ribose) Polymerase DNA-binding Domain: *trans*-dominant Inhibition of Poly(ADP-ribosyl)ation

To avoid in further studies the use of benzamides and any of their possible side effects (occurring especially at higher inhibitor concentration), we chose a molecular genetic approach to modulate poly(ADP-ribose) metabolism in living cells [19, 20]. We obtained a full-length PARP cDNA by library screening and reversed transcription/polymerase chain reaction. This cDNA and, in parallel, a fragment encoding only the DNA-binding domain of the enzyme were subcloned into eukaryotic expression vectors and transiently transfected into mammalian cells by electroporation. Transfected cells were ethanol-fixed, and PARP activity was reconstituted by postincubation with NAD, leading to the in situ production of poly(ADP-ribose). This was monitored by indirect immunofluorescence, using a monoclonal antibody (H10) directed against the polymer (kindly provided by H. Kawamitsu, Tokyo, Japan). Without NAD postincubation, there was a faint nuclear staining compared with the nonspecific cytoplasmic background (Fig. 4A). By contrast, NAD postincubation led to a brilliant nuclear staining, as is shown for vector-transfected control cells (Fig. 4B). Since transient transfection procedures reach only a fraction of cells (about 50 % in our experiments), it was important to identify transfected cells in situ. We therefore mixed antibody H10 with a rabbit antiserum against a part of the DNA-binding domain of PARP (FII-specific serum, kindly provided by G. de Murcia, Strasbourg, France) and performed double immunofluorescence studies. Transfection of an expression construct encoding the full-length PARP cDNA (pPARP62) led to a strong increase in the nuclear anti-PARP fluorescence (Fig. 4C), whereas control cultures transfected with vector carrying no insert gave a weak and homogeneous staining (data not shown). Overexpressing cells showed a drastic increase in immunoreactivity against poly(ADP-ribose) (Fig. 4D), as compared with neighboring nontransfected cells, confirming that our full-length cDNA was biologically active. By contrast, transfection of an expression construct selectively coding for the DNA-binding domain (pPARP51) led to strong protein signals in the nuclei of transfected cells (Fig. 4E) but reduced the intensity of polymer staining and hence PARP activity in the same nuclei drastically (Fig. 4F). Nontransfected cells in the same microscopic field exhibited full enzyme activity. The observed inhibition of poly(ADP-ribosyl)ation appears to be due to competition for DNA strand breaks which are necessary to activate the resident

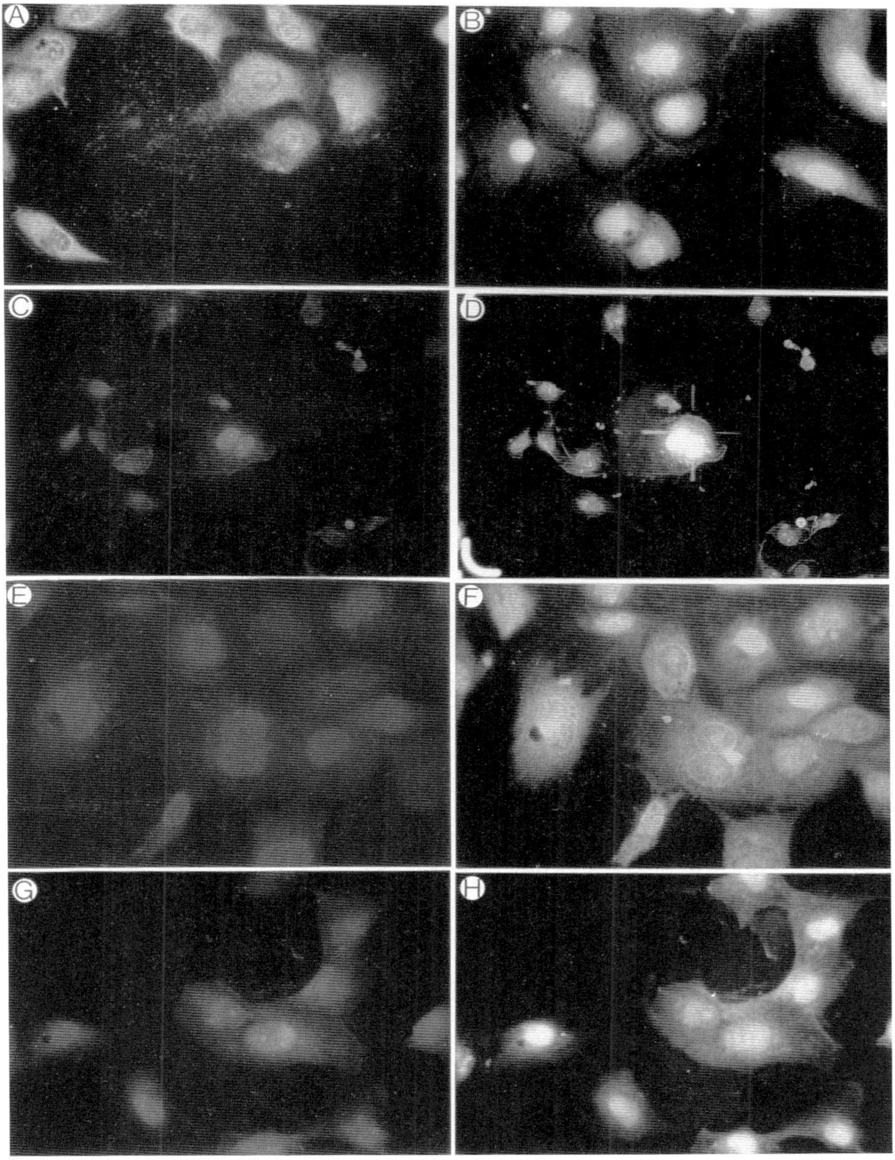

Fig. 4. Indirect double immunofluorescence of in vitro stimulated poly(ADP-ribose) Polymerase (PARP). Cells grown on coverslips were ethanol-fixed and postincubated with nicotinamide-adenine dinucleotide (NAD) (except in *A*); PARP is identified with tetramethylrhodamine isothiocyanate-conjugated goat anti-rabbit immunoglobulins (*C*, *E*, *G*); poly(ADP-ribose) is identified with fluorescein isothiocyanate-conjugated goat anti-mouse immunoglobulins (*A*, *B*, *D*, *F*, *H*); *A/B*, transfection of vector DNA lacking PARP sequences; *C/D*, transfection with pPARP62; *E/F*, transfection with pPARP51; *G/H*, transfection with pPARP51, ethanol-fixed cells were gamma-irradiated (240 Gy) before NAD postincubation. (For experimental details, see text and [19])

enzyme, since the effect can be overcome by excessive amounts of DNA strand breaks as induced by supralethal gamma-irradiation applied to ethanol-fixed cells before NAD postincubation (Fig. 4G/H).

Taken together, these results demonstrate that a selectively overexpressed PARP DNA-binding domain inhibits poly(ADP-ribosyl)ation catalyzed by the resident PARP enzyme in a *trans*-dominant fashion. Using such a strategy, we are now in the position to pursue studies on the biological role(s) of poly(ADP-ribosyl)ation in intact cells without application of chemical inhibitors, which have been widely used so far but unfortunately do have side effects.

References

1. Althaus FR, Richter C (1987) ADP-ribosylation of proteins. Enzymology and biological significance. (Molecular biology, biochemistry and biophysics 37) Springer, Berlin, Heidelberg, New York
2. Alitalo K, Schwab M (1986) Oncogene amplification in tumor cells. Adv Cancer Res 47: 235 – 281
3. Schimke RT (1988) Gene amplification in cultured cells. J Biol Chem 263: 5989 – 5992
4. Bürkle A (1989) Altern und genetische Instabilität. Futura 2: 19 – 22 (Zeitschrift des Boehringer Ingelheim Fonds)
5. Bürkle A, Meyer T, Hilz H, zur Hausen H (1987) Enhancement of *N*-methyl-*N'*-nitro-*N*-nitrosoguanidine-induced DNA amplification in a simian virus 40-transformed Chinese hamster cell line by 3-aminobenzamide. Cancer Res (1987) 47: 3632 – 3636
6. Bürkle A, zur Hausen H (1987) Influence of poly(ADP-ribose) metabolism on carcinogen-inducible DNA amplification. In: zur Hausen H, Schlehofer JR (eds) The role of DNA amplification in carcinogenesis. Lippincott, Philadelphia, pp 126 – 132
7. Bürkle A, Heilbronn R, zur Hausen H (1990) Potentiation of carcinogen-induced methotrexate resistance and dihydrofolate reductase gene amplification by inhibitors of poly(adenosine diphosphate-ribose) polymerase. Cancer Res 50: 5756 – 5760
8. Bürkle A (1989) Inhibition of carcinogen-inducible DNA amplification in a simian virus 40-transformed hamster cell line by ethacridine or ethanol. Cancer Res 49: 2584 – 2587
9. Shmookler-Reis RJ, Goldstein S (1980) Loss of reiterated DNA sequences during serial passage of human diploid fibroblasts. Cell 21: 739 – 749
10. Harley CB, Futcher AB, Greider CW (1990) Telomeres shorten during aging of human fibroblasts. Nature 345: 458 – 460
11. Lundblad V, Szostak JL (1989) A mutant with a defect in telomere elongation leads to senescence in yeast. Cell 57: 633 – 643
12. Kunisada T, Yamagishi H, Ogita Z-I, Kirakawa T, Mitsui Y (1985) Appearance of extrachromosomal circular DNAs during in vivo and in vitro aging of mammalian cells. Mech Aging Dev 29:89 – 99
13. Hart RW, Setlow RB (1974) Correlation between deoxyribonucleic acid excision-repair and life-span in a number of mammalian species. Proc Natl Acad Sci USA 71: 2169 – 2173
14. Hart RW, Sacher GA, Hoskins TL (1979) DNA repair in a short- and a long-lived rodent species. J Gerontol 34: 808 – 817
15. Francis AA, Lee WH, Regan JD (1981) The relationship of DNA excision repair of ultraviolet induced lesions to the maximum life span of mammals. Mech Aging Dev 16: 181 – 189
16. Pero RW, Holmgren K, Persson L (1985) Gamma-radiation induced ADP-ribosyl transferase activity and mammalian longevity. Mutat Res 142: 69 – 73

17. Berger NA, Petzold SI (1985) Identification of minimal size requirements of DNA for activation of poly(ADP-ribose) polymerase. Biochemistry 24: 4352 – 4355
18. Grube K, Küpper JH, Bürkle A (1991) Direct stimulation of poly(ADP-ribose) polymerase in permeabilized cells by double-stranded DNA oligomers. Anal Biochem 193: 236 – 239
19. Küpper JH, de Murcia G, Bürkle A (1990) Inhibition of poly(ADP-ribosyl)ation by overexpressing the poly(ADP-ribose) polymerase DNA-binding domain in mammalian cells. J Biol Chem 265: 18721 – 18724
20. Küpper JH (1990) Thesis, University of Heidelberg, FRG

Hormone Production in Neuroendocrine Tumours of Transgenic Mice*

G. Rindi, E. Solcia, and J.M. Polak

Introduction

Tumours producing regulatory peptides or amines common to the nervous and endocrine systems are called neuroendocrine tumours [1, 2]. They are composed of variably differentiated cells that share a number of morphologic markers normally expressed by neurons and endocrine cells [3, 4]. In laboratory animals, the occurrence of neuroendocrine tumours, which is virtually null in young animals, shows an extraordinary increase with age [5, 6]. The reasons for this phenomenon are still poorly understood; however it has been hypothesised that neuroendocrine tissues become particularly susceptible to transformation following age-related endocrinological imbalances [5 – 7].

Inducing neuroendocrine tumours in experimental animals could be of help in analysing the above phenomenon. Recent gene transfer techniques (the ability to transfer genetic information into the germ line of mammals) [8 – 10] made it possible to obtain laboratory animals developing neuroendocrine tumours at an early stage of life [11 – 20]. Inducing proliferation in cells specialised in the production of a given peptide (neuroendocrine cell) was achieved by using DNA constructs made up of the regulatory sequences of the hormone gene, aligned so as to transcribe sequences coding for transforming oncoprotein(s). The early region of the simian virus 40 (SV40), encoding the potent oncoprotein large T antigen [21, 22], has been largely employed for this purpose.

The complete characterisation of such tumours is the first and necessary step to enrol such animals as possible models of human disease. Here we report our experience in analysing transgenic models of neuroendocrine tumours.

---

*G.R. received a postdoctoral fellowship from IRCCS Policlinico San Matteo, Italy; this work was supported in part by the Italian Research Council (Oncology Project), Italy, and by the Cancer Research Campaign, UK.

Reihe der Villa Vigoni, Band 1
Biology of Aging
Zwilling and Balduini (Eds.)

Insulinomas in Insulin/SV40 Transgenic Mice

In this transgenic experiment D. Hanahan [12] used hybrid oncogenes composed of the rat insulin II gene 5'-flanking DNA linked to the SV40 early region (Tag). The resulting transgenic mice die prematurely while developing heritable tumours of the pancreas between 70 and 140 days of age. Sudden death was shown to be due to severe hypoglycemia, and they needed a lifelong hyperglucose diet to grow. Pancreas tumours were composed of proliferating insulin-producing B cells showing beta-granules on ultrastructural analysis [13]. In a subsequent analysis of these INS/SV40 tumours, Power et al. [14] identified a minor pancreatic polypeptide subpopulation in 30 % of examined growths. The whole endocrine pancreas was largely remodelled by the presence of hyperplastic islets. The oncoprotein large T antigen (Tag) was expressed in all islet B cells as well as in tumour cells. The development of pancreatic tumours seemed to result from Tag-promoted proliferation of B cells, and a sequence from hyperplasia to neoplasia was suggested [12].

This experiment provided evidence that 520 bp of the 5'-flanking rat insulin II gene DNA were capable of directing the expression of SV40 large T antigen to pancreatic insulin-producing B cells.

Pituitary Tumours and Insulinomas in Arginine-Vasopression/SV40 Transgenic Mice

This model was developed by Murphy et al. [16]. The transgene in use was made up of 1.25 kb of the 5'-flanking DNA deriving from the bovine arginine-vasopressin gene (AVP), linked to the early region of SV40. This transgenic experiment aimed at assessing the regulatory properties of the putative promoter of the vasopressin gene. Of six founder transgenic mice, one lineage (VT-C) was derived which showed a heritable pathology leading transgenic mice to death between 12 and 20 weeks. VT-C mice developed neuroendocrine tumours of the anterior pituitary and pancreas. None of these growths expressed arginine-vasopressin [16]. We carried out an extensive analysis of the VT-C mice phenotype [17]. Tumours of the pancreas were composed of insulin-producing B cells with a significant pancreatic polypeptide (PP)-immunoreactive subpopulation (in 25 % of tumours). Pituitary tumours were composed of poorly granulated cells which failed to react to the tests performed for the hormones normally produced in the anterior pituitary [16, 17]. By time course histological analysis [17], it was possible to establish that the endocrine pancreas underwent extensive remodelling which started early after birth (20 – 30 days of age). Dysplastic changes were atypical cells found either singly or organised into small groups inside well-formed islets. All atypical cells expressed insulin and were positive for large T antigen, the transforming oncoprotein encoded by the transgene. In the pituitaries similar lesions appeared later, with cytologic features of severe anaplasia, and the time which elapsed between the detection of early lesions and the development of large neoplasms was significantly shorter than in the endocrine pancreas. In addition, the hypothalamic vasopressinergic system of transgenic mice was unaffected. By immunohistochemistry study, no expression of the transgene was detected in those areas of the central nervous system where arginine-vasopressin is normally produced, neither was the synthesis of vasopressin perturbed [16].

We concluded that the transgene under study was capable of inducing transformation in two separate endocrine organs. The transgene-targeted cells were the insulin-producing B cells of the pancreas and an unspecified endocrine element of the anterior pituitary. Additionally, once initiated, the proliferative processes showed different paces in the two target organs.

The above transgenic experiment indicated that the vasopressin sequences used to build the transgene were not sufficient to direct the expression of the SV40 early region (Tag) to the hypothalamic vasopressinergic neurons. It was hypothesised that further regulatory sequences could be required for site-specific expression, as suggested by complex regulatory patterns elucidated in other transgenic experiments [15, 23 – 26]. Moreover, possible effects due to the chromosomal integration site could not be ruled out [16]. VT-C transgenic mice constitute an intriguing model for the study of multiple endocrine neoplasia syndrome type I (MEN I) [27, 28].

Glucagonomas in Glucagon/SV40 Transgenic Mice

S. Efrat et al. [18] produced these transgenic mice by using a hybrid oncogene made by the rat preproglucagon gene 5'-flanking DNA linked to the SV40 early region (Tag). Resulting GLU/Tag transgenic mice heritably developed tumours of the pancreas at 9 – 12 months of age. Pancreatic growths were isolated lesions composed of glucagon-producing A cells showing typical alpha granules at the ultrastructural level (Fig. 1) [29]. All tumour cells coexpressed large T antigen, as did proliferating A cells, forming peripheral mantles in rare hyperplastic islets. A sequence from focal hyperplasia to neoplasia was proposed to monitor tumour development in A cells. Tag expression was also detected in neurons of those areas of the central nervous system (CNS) where the preproglucagon gene is normally expressed and in other, unexpected sites of CNS. Unexpected Tag-positive neurons failed to colocalize glucagon or glucagon-like peptide (GLP) I, the parts of preproglucagon peptide for which they were tested.

It was concluded that 500 bp of the 5'-flanking rat preproglucagon gene DNA were capable of directing the expression of Tag site-specifically to pancreatic glucagon-producing A cells and to CNS neurons. The fact that, despite Tag expression, no appreciable proliferation of neurons could be identified was attributed to the transformation resistance properties inherent in neurons and their lack of proliferative capacity.

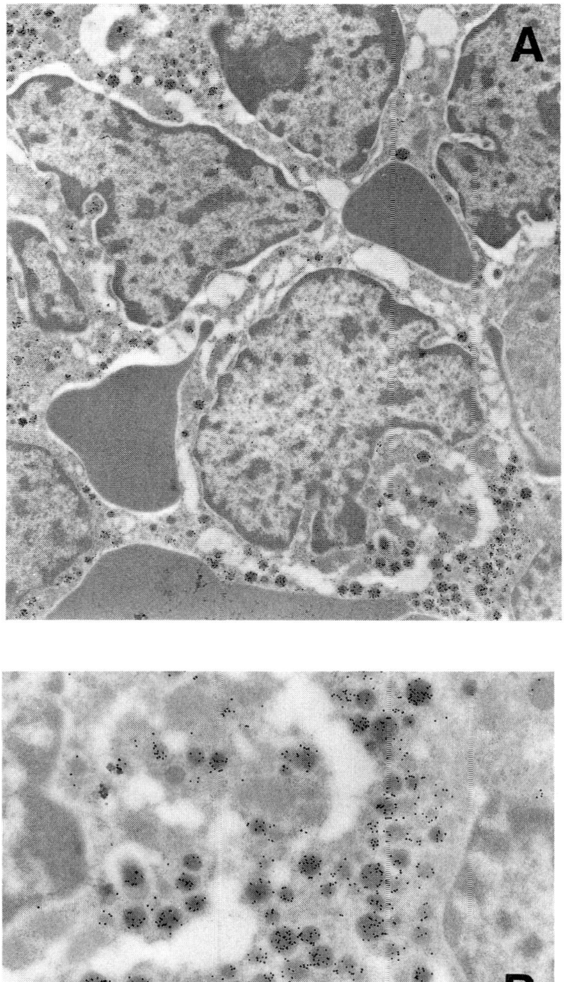

Fig. 1 A, B. Ultrastructure of GLU/Tag transgenic mouse pancreatic tumour. Tumour cells show numerous electron-dense endocrine granules. Immunological postembedding techniques allowed the detection of glucagon immunoreactivity in the endocrine granules. Immunogold labelling (10 nm diameter gold particles), uranyl acetate and lead citrate counterstain, × 15 000 (A), × 50 000 (B, which is detail of A)

Pancreatic Insulinomas and Intestinal Secretin-Producing Endocrine Tumours of Insulin-SV40/Insulin-PyST Transgenic Mice

Grant et al. [19, 20] developed transgenic mice carrying two hybrid oncogenes made up of the rat insulin II gene promoter (RIP) [12] aligned so as to transcribe either the early region of SV40 (Tag) or the polyomavirus small T antigen coding sequences (PyST) [21]. Double transgenic mice were derived by crossing RIP1/Tag2 mice [12] (see above) with RIP2/PyST1 mice. RIP2/PyST1 transgenic mice contained a transgene made with the insulin promoter (RIP) [12] linked to the polyomavirus small T antigen cDNA (PyST). This transgenic experiment was planned to study the possible effect of the polyoma oncoprotein small T antigen on B cells; however, no pathology resulted.

RIP1Tag2/RIP2PyST1 double transgenic mice developed insulinomas of the pancreas, as well as highly metastatic neuroendocrine tumours of the small intestine [19, 20]. Tumours of the pancreas retained the morphofunctional aspects previously described [12 – 14], being composed of insulin-producing B cells with a significant PP-immunoreactive subpopulation in 42 % of neoplasms analysed. Neuroendocrine tumours of the small intestine did not express insulin but were mainly composed of secretin-producing S-type cells [20]. Other gastrointestinal hormones (glicentin, gastrin, glucose-dependent insulinotropic polypeptide or GIP, neurotensin, somatostatin, and PP) were also detected in minor subpopulations [20]. In addition, by radioimmunoassay of plasma, high levels of insulin and secretin immunoreactants were detected, confirming the secretory properties of both pancreatic and intestinal growths [20]. SV40 large T antigen expression was detected in all B cells of non-tumour pancreatic islets and in some mucosal secretin cells of the small intestine [20]. It was concluded that the insulin promoter, besides inducing transformation in pancreatic B cells, is also capable of promoting tumour genesis in the intestinal secretin cell. This phenomenon suggests the existence of a possible relationship between B cells and S cells which underlies the capability of the transgene to express in both these sites.

The above transgenic model is a unique source of double hyperproduction of hormones and is the first recorded model for endocrine tumour genesis in mammalian intestine. Similar to INS/SV40 mice but at variance with the previously discussed VT-C mice, the progressive derangement of the endocrine pancreas embraced diffusely the entive B-cell population. The oncoprotein large T antigen was detected in almost all B cells, leading to a diffuse pattern of hyperplastic-dysplastic-adenomatous changes.

Conclusions

We have reviewed four transgenic mice models in which peptide-promoted oncoproteins were successfully used to induce oncogenesis in neuroendocrine cells. The unexpected transgene expression, which was demonstrated in some of the above experiments, suggests complex and possibly common regulatory patterns for hormone gene expression in neuroendocrine tissues. Transgenic mice heritably developing neuroendocrine tumours are therefore proposed as new tools for expanding our knowledge of gene expression in the neuroendocrine system. Finally, neuroendocrine tumours developing in transgenic mice show a number of morphofunctional features comparable with those found in corresponding human tumours and could form possible models of human disease.

References

1. Polak JM, Bloom SR (1985) Endocrine tumours. Churchill-Livingstone, Edinburgh
2. Solcia E, Rindi G, Capella C (1990) Neuroendocrine tumours and hyperplasias. In: Felipe MI, Lake BD (eds) Histochemistry in pathology. Churchill-Livingstone, Edinburgh, pp 397 – 409
3. Solcia E, Capella C, Buffa R, Tenti P, Rindi G, Corraggia M (1986) Antigenic markers of neuroendocrine tumours: their diagnostic and prognostic value. In: Fenoglio-Preiser CM, Weinstein RS, Kaufman N (eds) New concepts in neoplasia as applied to diagnostic pathology. Williams and Wilkins, Baltimore, pp 242 – 261 (International Academy of Pathology Monograph)
4. Bishop AE, Power RF, Polak JM (1988) Markers for neuroendocrine differentiation. Pathol Res Pract 183: 119 – 128
5. DeLellis RA, Nunnemacher G, Bitman WR, Gagel RF, Tashjian AH, Blount M, Wolfe HJ (1979) C cell hyperplasia and medullary carcinoma of the thyroid in the rat. Lab Invest 1979: 140 – 154
6. Lee AK, Ronald BS, DeLellis A, Blount M, Nunnemacher G, Wolfe HJ (1982) Pituitary proliferation lesion in aging male Long-Evans rats. Lab Invest 47: 595 – 602
7. Dihman VM (1971) Age associated elevation of hypothalamic threshold to feedback control and its role in development, aging and disease. Lancet I: 1211 – 1219
8. Gordon JW, Scangos GA, Plotkin DJ, Barbosa JA, Ruddle FH (1980) Genetic transformation of mouse embryos by microinjection of purified DNA. Proc Natl Acad Sci USA 77: 7380 – 7384
9. Hogan B, Costantini F, Lacy PE (1986) Manipulating the mouse embryo: a laboratory manual. Cold Spring Harbor, New York
10. Murphy D and Hanson J (1988) The production of transgenic mice by the microinjection of fertilized one-cell eggs. In: Glover DM (ed) DNA cloning: a practical approach, IRL, London, pp 213 – 248
11. Rindi G, Solcia E, Polak JM (1989) Transgenic mouse models and peptide producing endocrine tumours: morphofunctional aspects. In: Polak JM (ed) Regulatory peptides. Birkhäuser, Basel, pp 210 – 219
12. Hanahan D (1985) Heritable formation of pancreatic B-cell tumours in transgenic mice expressing recombinat insulin simian virus 40 oncogenes. Nature 315: 115 – 122
13. Holm RF, Varndell I, Power RF et al (1988) Ultrastructure and electron immunocytochemistry of insulin-producing B-cell tumours from transgenic mice: comparison with counterpart human tumour. Ultrastruct Pathol, 12: 547 – 559
14. Power RF, Holm RF, Bishop AE et al (1987) Transgenic mouse model: a new approach for the investigation of endocrine pancreatic B-cell growth. Gut 28: 121 – 129
15. Messing A, Chen HY, Palmiter RD, Brinster RL (1985) Peripheral neuropathies, hepatocellular carcinomas and islet cell adenomas in transgenic mice. Nature 316: 461 – 463
16. Murphy D, Bishop AE, Rindi G et al (1987) Mice transgenic for a vasopressin hybrid oncogene develop tumours of the endocrine pancreas and the anterior pituitary. A possible model for human multiple endocrine neoplasia syndrome type I. Am J Pathol 129: 552 – 566

17. Rindi G, Bishop AE, Murphy D, Solcia E, Hogan B, Polak JM (1988) A morphological analysis of endocrine tumour genesis in pancreas and anterior pituitary of AVP/SV40 transgenic mice. Virchows Arch A 412: 255 – 266
18. Efrat S, Teitelman G, Anwar M, Ruggiero D, Hanahan D (1988) Glucagon gene regulatory region directs oncoprotein expression to neurons and pancreatic alpha-cells. Neuron 1: 605 – 613
19. Grant SGN, Seidman I, Hanahan D, Bautch VL (1991) Early invasiveness characterises metastatic carcinoid tumours in transgenic mice. Lancer Research 51: 4917 – 4923
20. Rindi G, Grant SGN, Yiangou Y et al (1990) Development of neuroendocrine tumours in the gastrointestinal tract of transgenic mice: heterogeneity of hormone expression. Am J Pathol 136: 1349 – 1363
21. Tooze J (1982) Molecular biology of tumour viruses, part 2: DNA tumour viruses, 2nd edn. Cold Spring Harbor, New York
22. Rigby PWJ, Lane DE (1983) Structure and function of simian virus 40 large T-antigen. In: Klein G (ed) Advances in viral oncology. Raven, New York, pp 31 – 57
23. Palmiter RD, Brinster RL, Hammer RE, Trumbauer ME, Rosenfeld MG, Birnberg NC, Evans RM (1982) Dramatic growth of mice that develop from eggs microinjected with metallothionein growth hormone fusion genes. Nature 300: 611 – 615
24. Hammer RE, Krumlauf R, Camper S, Brinster RL, Tilghman JM (1987) Diversity of alpha-fetoprotein gene expression is generated by a combination of separate enhancer elements. Science 235: 53 – 58
25. Krumlauf R, Hammer RE, Brinster RL, Chapman VM, Tilghman JM (1985) Regulated expression of alpha-fetoprotein genes in transgenic mice. Cold Spring Harbor Symp Quant Biol 316: 371 – 378
26. Palmiter RD, Chen HY, Messing H, Brinster RL (1985) SV40 enhancer and large T-antigen are instrumental in the development of choroid tumours in transgenic mice. Nature 316: 457 – 460
27. Wermer P (1954) Genetic aspects of adenomatosis of endocrine glands. Am J Med 16: 363 – 371
28. De Lellis RA, Dayal Y, Tischler AS, Lee AK, Wolfe JH (1986) Multiple endocrine neoplasis (MEN) syndromes: cellular origins and interrelationships. Int Rev Exp Pathol 28: 163 – 215
29. Rindi G, Efrat S, Ghatei MA, Bloom SR, Solcia E, Polak JM (1991) Glucagonomas of transgenic mice express a wide range of general neuroendocrine markers and bioactive peptides. Virchows Arch A 419: 115 – 129

Genetic Control of Aging in the Ascomycete *Podospora anserina**

H.D. Osiewacz

Introduction

Podospora anserina is a filamentous fungus which shows after prolonged vegetative growth different symptoms of aging: The growth rate of a colony (mycelium) decreases, the pigmentation of the culture changes, the peripheral hyphae (filamentous "cells" of a mycelium) become slender and undulate, and the formation of aerial hyphae is drastical reduced. Some days after the first occurrence of these morphological symptoms of aging, the culture stops growing and dies. Since the first description of this "senescence syndrome" in the early 1950s [37], senescence has been extensively investigated at the physiological, genetic, and molecular levels. Emerging from these investigations are first ideas about the molecular mechanisms controlling aging in this simple eukaryotic microorganism.

Several characteristics make this organism ideal for the experimental analysis of aging processes. (a) Cultures have a simple organization consisting of long highly branched cells (hyphae) which in total form a mycelium. Specialized cells (e.g., male gametes, spermatia) and "organs" (e.g., fruit bodies: perithecia) are only formed during sexual reproduction. (b) *P. anserina* can easily be cultured both in liquid as well as on solid medium. (c) The life spans of all wild strains are short (e.g., of wild-type strain s, 25 days). (d) Completion of the sexual cycle takes about 10 – 12 days. As a consequence generation times are short. (e) *P. anserina* is accessible to formal classical genetic analysis. After crossing two strains of opposite mating type, the products of meiosis can easily be isolated from an ascus (meiosporangium) which contains four linearly arranged spores (ascospores). Germinated ascospores give rise to new mycelia, which can be further analyzed. These characteristics allow the isolation and analyses of a large number of progeny of well-defined genetic crosses in a short time. (f) *P. anserina* is accessible to molecular genetics. Nucleic acids and proteins can be isolated from fungal cultures and can be analyzed by standard molecular techniques. In the past decade methods for the introduction of foreign DNA into *P. anserina* cells via transformation of protoplasts have been

*The experimental work was supported in part by a grant of the Deutsche Forschungsgemeinschaft (Bonn Bad-Godesberg).

Reihe der Villa Vigoni, Band 1
Biology of Aging
Zwilling and Balduini (Eds.)

developed [2, 7, 17, 33]. These techniques now allow the genetic manipulation of strains and in principle the isolation of well-defined nuclear genes from the genome of this ascomycete.

In this paper, recent investigations to elucidate the control of aging in *P. anserina* are summarized and discussed. A detailed description of the life cycle and genetics of *P. anserina* can be found in Esser and Kuenen [14] and Esser [11]. For previous work performed with *P. anserina* the reader is referred to Esser and Tudzynski [15] and Esser and Böckelmann [12].

Nuclear Genetic Factors Controlling the Life Span

The genetic analysis of a large number of different morphological mutants of *P. anserina* revealed that many are pleiotropic; not only morphological characteristics (e.g., color of the mycelium, growth rate) are affected but also the life span. The corresponding mutations were mapped to seven different linkage groups (for review, see Esser [11]). During this course of investigation, two double mutants (*gr viv* and *i viv*) derived from crosses between single nuclear mutants (*gr, viv*, and *i*) were isolated [13, 44]. These mutants phenotypically differ from their parents and do not show the characteristic symptoms of aging even after extended periods of vegetative growth. Thus, senescence appears to be prevented by the synergistic action of two single nuclear mutations. A detailed molecular analysis of these double mutants revealed that the liberation and/or amplification of a mitochondrial (mt) genetic trait is affected (see below).

Approaches to Isolate Nuclear Age-Related Genes

Although a large number of nuclear genes affecting the life span of *P. anserina* have been identified and characterized by formal genetic methods, not a single nuclear gene of this type has been cloned until now. A major problem in the isolation of specific genes from *P. anserina* is the fact that no efficient transformation system is available which allows a successful use of complex genomic libraries in complementation experiments.

In order to establish an experimental approach of this type, a generally applicable transformation system for *P. anserina* which can be used to transform wild-type strains and mutants has been developed. This system is based on the use of a plasmid vector which carries the dominant hygromycin B (*hph*) resistance gene of *Escherichia coli* under the control of regulation sequences derived from the closely related ascomycete *Aspergillus nidulans* [34]. Using this system different *P. anserina* strains (wild-type ones and mutants) have been transformed with efficiencies greater than 20 transformants per μg plasmid DNA [33].

As may be seen from Fig. 1, the wild-type strain s, used as the recipient in transformation experiments, does not contain sequences which hybridize to

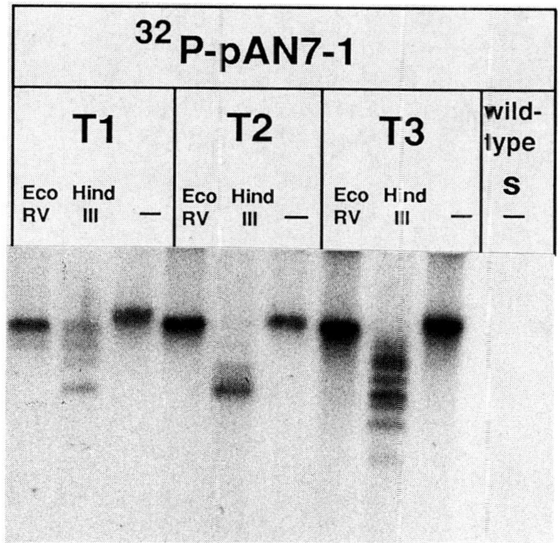

Fig. 1. Southern blot hybridization analyses of *Podospora anserina* wild-type strain s and three independent transformants (*T1 – T3*). Uncut (–), *Hind*III and *Eco*RV digested DNA from the wild-type recipient and from hygromycin B-resistant transformants were fractionated by agarose gel electrophoresis, blotted to nitrocellulose and hybridized to nick-translated pAN7-1 (data from [33])

the transforming vector. After transformation with vector pAN7-1, sequences of the transforming plasmid can clearly be detected in the total genomic DNA of hygromycin B-resistant transformants.

The data obtained from the hybridization analysis clearly revealed that the transforming vector becomes integrated into the genome of the recipient either as a single copy or as multiple, tandemly arranged arrays. Integration occurs at different positions in independent transformants. No hybridization of low molecular weight DNA has been obtained with uncut DNA of different transformants. Thus, no hints for autonomously replicating vector sequences have been provided by the molecular analysis of hygromycin B-resistant colonies.

Currently, modifications of the transformation procedure and the transformation vector aimed to improve efficiencies are in progress.

In a second approach to establish a method for the isolation of nuclear age-related genes of *P. anserina*, experiments were started to separate and characterize intact chromosomes by pulsed field gel electrophoresis. Using this method, it was possible to separate five chromosomal bands (Fig. 2). According to the relative fluorescence of the different bands, two chromosomes appear to comigrate in bands II and III, respectively. Thus, seven chromosomes can be identified on pulsed field gels. These molecular data exactly correspond to the number of genetically characterized linkage groups (for review, see Esser [11]).

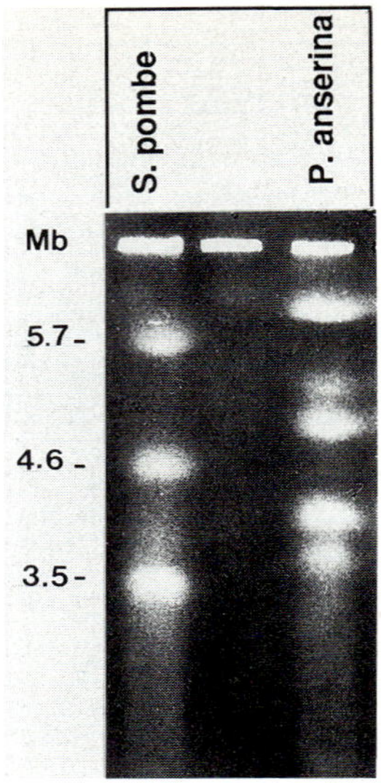

Fig. 2. Separation of *P.anserina* chromosomes by pulsed field gel electrophoresis. Chromosomal DNA from *P.anserina* was prepared and fractionated on a 0.7 % agarose gel as described [32]. Chromosomes of *Schizosaccharomyces pombe* were used as size markers. The approximate length of the separated chromosomes is indicated in megabases (Mb)

A combination of genetical (linkage analysis) and molecular (hybridization analysis) experiments resulted in the identification of the mating type locus on chromosome VII [33] and the rDNA gene unit on a chromosomal band migrating at about 4.9 Mb [32]. More detailed analyses of fractionated chromosomes using various probes will lead to a molecular characterization of the individual chromosomes. These data will be essential for the selection and recovery of individual chromosomes from pulsed field gels. Using these well-characterized fractions of the *Podospora* genome it should be possible to construct chromosome-specific cosmid libraries. Because of the lower complexity of these libraries, the isolation of nuclear genes by complementation of mutants should be possible, even though the transformation efficiencies are rather low. A subsequent characterization of age-related nuclear genes is expected to provide further clues to understanding the genetic control of aging in *P.anserina* and the molecular basis of aging processes in general.

Extrachromosomal Genetic Factors Correlate with the Onset of Senescence

From the genetic analyses of different wild-type strains and of different long-lived mutants, it has become obvious that the onset of senescence in *P.anserina* is not only controlled by nuclear genetic factors but also by extrachromosomal genetic traits [27]. Subsequently, the comparative molecular analysis of juvenile and senescent *Podospora* cultures revealed an additional DNA species of circular structure in senescent mitochondria [8, 40, 41]. In juvenile mycelia this DNA (termed plDNA or α-senDNA) is an integral part of the mtDNA and represents the first intron of the gene coding for cytochrome *c* oxidase subunit I (COI) [3, 9, 23, 30]. During aging the intron becomes liberated and amplified and can be isolated as a covalently closed circular DNA.

The crucial role of plDNA in the control of aging is demonstrated by a number of different results derived from the molecular analyses of both wild-type strains and long-lived mutants. In the next part of this chapter these data, which have led to the proposal of a model explaining the control of the aging process in *P. anserina* at the molecular level, will be summarized and discussed.

Molecular Analysis of Wild-Type Strains

The genetic and molecular analysis of various wild-type strains isolated from different parts of France and Germany revealed that they can be distinguished at the genetic as well as at the molecular level. These individual strains are characterized by their heterogenic incompatibility reaction in the sexual and/or the vegetative phase [24].

An analysis of their mtDNA revealed restriction fragment length polymorphisms (RFLP) between the different strains due to insertions and deletions in the mt genome, some of which could clearly be located by restriction enzyme mapping [24]. Recently, a more detailed comparison of the two wild-type races A and s revealed the presence of three introns in race A which are absent from the mtDNA of race s [10]. These introns are located in three different genes (ND4l, COI, Cytb) of the mt genome.

In contrast to these three introns, the first intron of the COI gene, which gives rise to the liberation and amplification of plDNA, appears not to be optional. All different wild isolates analyzed so far which show the normal symptoms of aging have been found to contain both the integrated copy (the intron) as well as the free circular molecule (plDNA). Thus, in contrast to other introns of *P. anserina* and of other organisms (e.g., *Saccharomyces cerevisiae*), this particular intron appears to have an essential function in wild-type strains.

Molecular Analysis of Long-Lived Mutants

A number of long-lived mutants have been isolated from *P. anserina* which subsequently have been analyzed by genetic as well as by molecular approaches. In a number of mutants the site of mutation has clearly been mapped to one of seven nuclear linkage groups. Another group of mutations has been shown to be maternally inherited, indicating extranuclear genetic traits to be the site of mutation. Among these different mutants there is a group which seems to have acquired eternal life (e.g. *ex1, ex2, i viv, gr viv*). On the other hand, mutants with reduced as well as increased life spans are known (e.g., *i, viv, z*).

The molecular analysis of the various long-lived mutants revealed either they do not contain amplified plDNA or the amplification of plDNA is delayed. In the nuclear double mutants *gr viv* and *i viv*, the liberation and amplification of plDNA is affected by the synergistic action of two nuclear genes [43]. In the two mt mutants *ex1* and *ex2*, no source for the formation of free plDNA is present because in these mutants most parts of the COI gene including the first intron are deleted [25, 38].

In mutant AL2, which is characterized by an increased life span (about 200 days vs. 15 days of the wild-type strain A), the site of mutation is located in the mtDNA but outside the COI gene (Fig. 3). Southern blot experiments revealed that in this mutant the amplification of plDNA is delayed. Most interestingly, a linear DNA species has been identified in mutant AL2 which shares characteristics with linear plasmids [31]. This group of genetic elements has been found in various organisms including filamentous fungi, yeasts, and higher plants, and some of them have been demonstrated to be correlated with particular phenotypes (for review, see [28]). In certain strains of the ascomycete *Neurospora intermedia*, the onset of senescence is correlated with the integration of linear plasmids into mtDNA and the accumulation of defective mt genomes [5, 6].

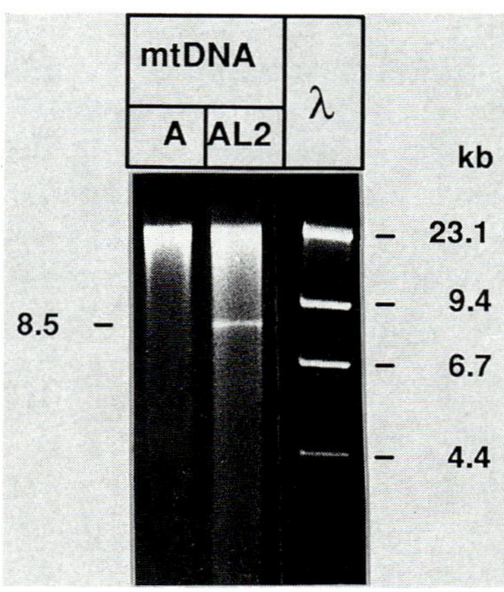

Fig. 3. Comparison of the mitochondrial (mt)DNA fraction from *P. anserina* wild-type strain race A and long-lived mutant AL2. Undigested DNA was fractionated on an agarose gel. An additional DNA species (pAL2-1) of about 8.5 kb is present in the DNA preparation of the mutant but absent from the wild-type strain

A Molecular Model Explaining the Genetic Control of Aging

From the compilation of the different data derived from the analyses of various wild-type strains and of long-lived mutants a model explaining the control of aging in *P. anserina* has been proposed [25]. In Figure 4 modification of this model is shown, in which results of recent investigations are integrated. According to the scenario shown in this scheme, the free amplified plasmid (plDNA) found in senescent *Podospora* cultures may be derived either from a direct excision at the DNA level (DNA excision) or from reverse transcription using the COI-specific pre-mRNA as a template. A polypeptide encoded by a long open reading frame (ORF) which is present on the first COI intron may be involved in the process of reverse transcription because the amino acid sequence deduced from the ORF shows significant homology to various well-characterized reverse transcriptases [29]. However, although a reverse transcriptase activity has been identified in crude extracts of middle-aged *Podospora* cultures and not in juvenile mycelia [42] the expression of the plDNA-encoded ORF and a reverse transcriptase activity of the corresponding polypeptide still remain to be demonstrated.

The data derived from the molecular analysis of the long-lived nuclear mutants *gr viv* and *i viv* revealed that nuclear genes may prevent the liberation and/or the amplification of plDNA. On the one hand, in *gr viv* free amplified plDNA has never been identified, indicating a defect in its liberation [44]. On the other hand, only small amounts of free plDNA have been detected in *i viv*, suggesting that the liberation mechanism and/or the amplification of plDNA is affected by the two mutated nuclear genes [44].

As indicated in Fig. 4, the role of plDNA in the control of senescence may be explained in different ways:

1. The plDNA may act as a mutator of mtDNA. After integration into essential genes the corresponding mutated defective genomes may accumulate, leading to deficiencies of essential mt gene products. In fact, a decrease of cytochrome *c* oxidase in senescent *Podospora* cultures has been described [4]. Deficiencies of this and other enzymes lead to a defective respiration chain and, as a consequence, cellular death.

2. The plDNA may integrate into mtDNA in one or multiple copies. In a subsequent recombination process between homologous sequences (resident and novel integrated plDNA copies), mt subcircles may be formed. Depending on whether these subcircles contain a functional origin of replication, they are retained in senescent cultures or become lost. This mechanism is consistent with the loss of large parts of the juvenile mtDNA found in senescent mycelia [23]. From more recent results, it has become clear that the deleted regions of mtDNA contain a large number of essential mt genes [10].

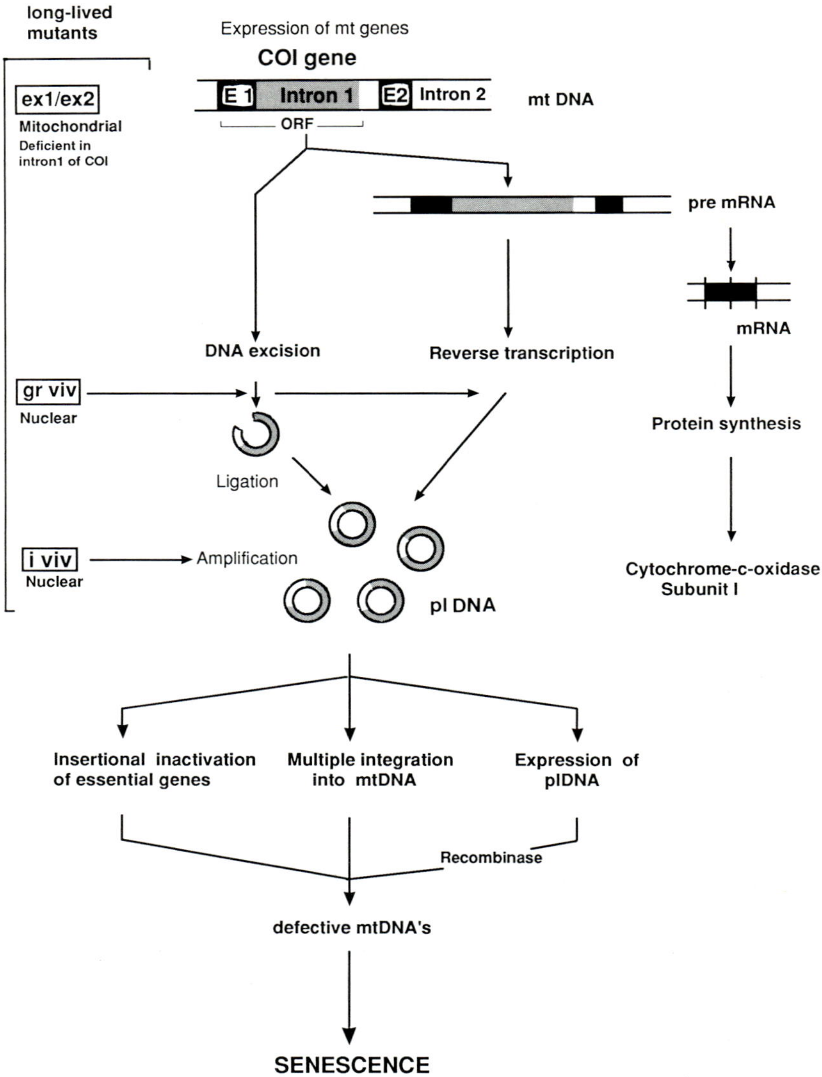

Fig. 4. Molecular model explaining the control of senescence in *P. anserina* (modified from [25]). In the mitochondrial (mt) gene coding for cytochrome *c* oxidase subunit I (COI) the first two exons (*E1*, *E2*, *black*) and the first two introns are indicated. The open reading frame (*ORF*) of intron 1 is designated by the *hatched area*. This ORF is also indicated in the pre-mRNA of the COI gene and the liberated intron sequence (*plDNA*). For details see text

3. The ORF of the amplified plDNA may become overexpressed, leading to the presence of a polypeptide with recombinase activity in senescent mycelia. The action of this enzyme may explain the formation and accumulation of defective mt genomes found in senescent *Podospora* cultures. Most interestingly, a recombinase activity has been demonstrated in *E.coli* for the fourth intron (bI4) of the apocytochrome B gene (Cytb) of *S.cerevisiae* [19]. In *P.anserina* the data derived from the characterization of the long-lived mutants *ex1* and *ex2* are consistent with a recombinase activity encoded by plDNA. In these mutants the mtDNA is "stabilized" after deletion of large parts of the COI gene including the first intron [38].

Significance of Data Obtained with *P.anserina* for Understanding the Control of Aging Processes in Higher Organisms

From the investigations performed with *P.anserina* in the past three decades, the control of senescence by genetic factors has clearly been demonstrated. After a correlation of the amplification of a mobile genetic element (plDNA) and the control of the onset of senescence was shown in this ascomycete, very similar genetic elements were identified in other closely related fungi (for review, see [16, 28]). In all cases in which these elements have been coupled to control of the aging process, they have been shown to be involved in rearrangements of the mt genome and thus appear to function as mutators [1, 5]. As a result, the inactivation of particular essential mt functions (e.g, components of the respiration chain) may finally lead to cellular death.

Most interestingly, investigations over the past several years clearly point to mitochondria as cell compartments involved in the control of aging processes in humans. These investigations have revealed point mutations as well as deletions in mtDNA of human muscle cells of patients with mitochondrial myopathies (e.g, Kearns – Sayre syndrome) [20, 45, 47]. In some cases deficiencies of components of the respiration chain are the consequence of these mutations. In addition, respiratory-defective cells have been detected in the heart and skeletal muscle of healthy persons (for review, see [22]). Finally, very recently, an age-related accumulation of mt genomes has been identified in cells in the striatum of normal persons as well as in patients with Parkinsons disease [21]. These data point to a crucial role of mitochondria in the control of aging also in organisms of higher organization [22].

As postulated by Kadenbach and Müller-Höcker (1990), the life span in humans may be limited by a continuous accumulation of respiratory-deficient cells, mainly in the heart. However, it is still not clear whether the mtDNA mutations identified in various human cells are the consequence of statistic mutations or whether they are controlled by genetic traits. Like in fungi, nuclear factors may be involved in this control. In this respect, it is noteworthy that nuclear DNA rearrangements have been shown to be associated with aging in different mammalian systems (for review, see [39]). In addition, extrachromosomal circular molecules have been identified in various eukaryotic tissues and cell culture systems (e.g. human fibroblasts) [26, 35, 36, 46]. Some of these elements share characteristics of typical transposons [18] and may be correlated with nuclear DNA rearrangements that occur during aging. As a consequence, factors involved in controlling the stability of the mt genetic system may become defective, leading to those mutations of mt genomes which have been identified in the different cell types. Thus, although the mechanisms controlling aging processes in various organisms may differ in detail from each other, it is possible that the basic mechanisms leading to cellular death may be found to be comparable.

Acknowledgements: I am grateful to Prof. Harald zur Hausen (Heidelberg) for his interest and valuable support and to Dr. Rüdiger Ridder (Heidelberg) for critical reading of the manuscript.

References

1. Akins RA, Kelley RL, Lambowitz AM (1986) Mitochondrial plasmids of *Neurospora*: integration into mitochondrial DNA and evidence for reverse transcription in mitochondria. Cell 47: 505 – 516
2. Bégueret J, Razanamparany V, Perrot M, Barreau C (1984) Cloning gene *ura5* for the orotidylic acid pyrophosphorylase of the filamentous fungus *Podospora anserina*: transformation of protoplasts. Gene 32: 487 – 492
3. Belcour L, Begel O, Mosse MO, Vierny C (1981) Mitochondrial DNA amplification in senescent cultures of *Podospora anserina*. Curr Genet 3: 13 – 21
4. Belcour L, Begel O, Keller AM, Vierny C (1982) Does senescence in *Podospora anserina* result from instability of the mitochondrial genome. In: Slonimski PP, Borst P, Attardi G (eds) Mitochondrial genes. Cold Spring Habor, New York, pp 415 – 422
5. Bertrand H, Chan BSS, Griffiths AIF (1985) Insertion of a foreign nucleotide sequence into mitochondrial DNA causes senescence in *Neurospora intermedia*. Cell 41: 877 – 884
6. Bertrand H, Griffiths AJF, Court DA, Cheng CK (1986) An extrachromosomal plasmid is the etiological precursor of kalDNA insertion sequences in the mitochondrial chromosome of senescent *Neurospora*. Cell 47: 829 – 837
7. Brygoo Y, Debuchy R (1985) Transformation by integration in *Podospora anserina*: I. Methodology and phenomenology. Mol Gen Genet 200: 128 – 131
8. Cummings DJ, Belcour L, Grandchamps C (1979) Mitochondrial DNA of *Podospora anserina*: II. Properties of mutant DNA and multimeric circular DNA from senescent cultures. Mol Gen Genet 171: 239 – 250
9. Cummings DJ, MacNeil IA, Domenico J, Matsuura ET (1985) Excision-amplification of mitochondrial DNA of *Podospora anserina*. DNA sequence analysis of three unique 'plasmids'. J Mol Biol 185: 659 – 680
10. Cummings DJ, McNally KL, Domenico JM, Matsuura ET (1990) The complete DNA sequence of the mitochondrial genome of *Podospora anserina*. Curr Genet 17: 375 – 402
11. Esser K (1974) *Podospora anserina*. In: King RC (ed) Handbook of genetics, vol I. Plenum, New York, pp 531 – 551
12. Esser K, Böckelmann B (1985) Fungi. In: Lints FA (ed) Nonmammalian models for research on aging. Interdisciplinary topics in gerontology, vol 21. Karger, Basel, pp 231 – 246
13. Esser K, Keller E (1976) Genes inhibiting senescence in the ascomycete *Podospora anserina*. Mol Gen Genet 144: 107 – 110

14. Esser K, Kuenen R (1967) Genetics of fungi. Springer, Berlin, Heidelberg, New York
15. Esser K, Tudzynski P (1980) Senescence in fungi. In: Thimann KV (ed) Senescence in plants. CRC, Boca Raton, pp 67 – 83
16. Esser K, Kück U, Lang-Hinrichs C, Lemke P, Osiewacz HD, Stahl U, Tudzynski P (1986) Plasmids of eukaryotes. Springer, Berlin, Heidelberg, New York
17. Fernandez-Larrea J, Stahl U (1989) Transformation of *Podospora anserina* with a dominant resistance gene. Curr Genet 16: 57 – 60
18. Fujimoto S, Yamagishi H (1987) Isolation of an excision product of T-cell receptor α-chain gene rearrangements. Nature 327: 242 – 243
19. Goguel V, Bailone A, Devoret R, Jacq C (1989) The bI4 RNA mitochondrial maturase of *Saccharomyces cerevisiae* can stimulate intra-chromosomal recombination in *Escherichia coli*. Mol Gen Genet 216: 70 – 74
20. Holt JL, Harding AE, Morgan-Hughes JA (1988) Deletions of muscle mitochondrial DNA in patients with mitochondrial myopathies. Nature 331: 717 – 719
21. Ikebe S, Tanaka M, Ohno K, Sato W, Hattori K, Kondo T, Mizuno Y, Ozawa T (1990) Increase of deleted mitochondrial DNA in the striatum in Parkinson's disease and senescence. Biochem Biophys Res Commun 170: 1044 – 1048
22. Kadenbach B, Müller-Höcker J (1990) Mutations of mitochondrial DNA and human death. Naturwissenschaften 77: 221 – 225
23. Kück U, Stahl U, Esser K (1981) Plasmid-like DNA is part of the mitochondrial DNA in *Podospora anserina*. Curr Genet 5: 143 – 147
24. Kück U, Kappelhoff B, Esser K (1985 a) Despite mtDNA polymorphism the mobile intron (plDNA) of the COI gene is present in different races of *Podospora anserina*. Curr Genet 10: 59 – 67
25. Kück U, Osiewacz HD, Schmidt B, Kappelhoff B, Schulte E, Stahl U, Esser K (1985 b) The onset of senescence is affected by DNA rearrangements of a discontinuous mitochondrial gene in *Podospora anserina*. Curr Genet 9: 373 – 382
26. Kunisada T, Yamagishi H, Ogita Z, Hirakawa T, Mitsui Y (1985) Appearance of extrachromosomal circular DNAs during *in* vivo and *in* vitro aging of mammalian cells. Mech Aging Dev 29: 89 – 99
27. Marcou D (1961) Notion de longévité et nature cytoplasmatique du déterminant de la sénescence chez quelques champignons. Ann Sci Nat Bot 12: 653 – 764
28. Meinhardt F, Kempken F, Kämper J, Esser K (1990) Linear plasmids among eukaryotes: fundamentals and application. Curr Genet 17: 89 – 95
29. Michel F, Lang B (1985) Mitochondrial class II introns encode proteins related to the reverse transcriptases of retroviruses. Nature 316: 641 – 643
30. Osiewacz HD, Esser K (1984) The mitochondrial plasmid of *Podospora anserina*: a mobile intron of a mitochondrial gene. Curr Genet 8: 299 – 305
31. Osiewacz HD, Hermanns J, Marcou D, Triffi M, Esser K (1989) Mitochondrial DNA rearrangements are correlated with a delayed amplification of the mobile intron (plDNA) in a long-lived mutant of *Podospora anserina*. Mutat Res 219: 9 – 15
32. Osiewacz HD, Clairmont A, Huth M (1990) Electrophoretic karyotype of the ascomycete *Podospora anserina*. Curr Genet 18: 481 – 483
33. Osiewacz HD, Skaletz A, Esser K (1991) Integrative transformation of the ascomycete *Podospora anserina*: identification of the mating type locus on chromosome VII of electrophoretically separated chromosomes. Appl Microbiol Biotechnol 35: 38 – 45
34. Punt PJ, Oliver RP, Dingenmanse MA, Pouwels PH, van den Hondel CAMJJ (1987) Transformation of *Aspergillus* based on the hygromycin B resistance marker from *Escherichia coli*. Gene 56: 117 – 124
35. Riabowol KT, Shmookler Reis RJ, Goldstein S (1985 a) Properties of extrachromosomal covalently closed circular DNA isolated and cloned from aged human fibroblasts. Age 8: 114 – 121
36. Riabowol KT, Shmookler Reis RJ, Goldstein S (1985 b) Tandemly repetitive and interspersed repetitive sequences are differentially represented in extrachromosomal covalently closed circular DNA of human diploid fibroblasts. Nucleic Acids Res 13: 5563 – 5584

37. Rizet G (1953) Sur l'impossibilité d'obtenir la multiplication vegetative ininterrompue et illimité de l'ascomycete *Podospora anserina*. C R Acad Sci Paris 237: 838 – 855
38. Schulte E, Kück U, Esser K (1988) Extrachromosomal mutants of *Podospora anserina*: permanent vegetative growth in spite of multiple recombination events in the mitochondrial genome. Mol Gen Genet 211: 342 – 349
39. Slagboom PE, Vijg J (1989) Genetic instability and aging: theories, facts, and future perspectives. Genome 31: 373 – 385
40. Stahl U, Lemke PA, Tudzynski P, Kück U, Esser K (1978) Evidence for plasmid like DNA in a filamentous fungus, the ascomycete *Podospora anserina*. Mol Gen Genet 162: 341 – 343
41. Stahl U, Kück U, Tudzynski P, Esser K (1980) Characterization and cloning of plasmid like DNA of the ascomycete *Podospora anserina*. Mol Gen Genet 178: 639 – 646
42. Steinhilber W, Cummings DJ (1986) A DNA polymerase activity with characteristics of a reverse transcriptase in *Podospora anserina*. Curr Genet 10: 389 – 392
43. Tudzynski P, Esser K (1979) Chromosomal and extrachromosomal control of senescence in the ascomycete *Podospora anserina*. Mol Gen Genet 173: 71 – 84
44. Tudzynski P, Stahl U, Esser K (1982) Development of a eukaryotic cloning system in *Podospora anserina*: I. Long-lived mutants as potential recipients. Curr Genet 6: 219 – 222
45. Wallace DC (1989) Mitochondrial DNA mutations and neuromuscular disease. TIBS 5: 9 – 13
46. Yamagishi H, Kunisada T, Takeda T (1985) Amplification of extrachromosomal small circular DNAs in a murine model of accelerated senescence. A brief note. Mech Aging Dev 29: 101 – 103
47. Zeviani M, Servidei S, Gellera C, Bertini E, DiMauro S, DiDonato S (1989) An autosomal dominant disorder with multiple deletions of mitochondrial DNA starting at the D-loop region. Nature 339: 309 – 311

Subject Index

Printing: Druckerei Kutschbach, Berlin
Binding: Buchbinderei Lüderitz & Bauer, Berlin